# Vespa Scooters 90,125,150, 18 and 200cc Owners Workshop Manual

## by Jeff Clew

**Models covered:**

| | | |
|---|---|---|
| V9A1 | 88.5 cc | Vespa 90 |
| V9SS1 | 88.5 cc | Vespa 90 Super Sport |
| V9SS2 | 88.5 cc | Racer |
| 232L2 | 123.4 cc | |
| VMA1 | 121.1 cc | Vespa 125 |
| VMA2 | 121.1 cc | Primavera |
| 312L2 | 145.45 cc | Sportique |
| VBC1 | 145.45 cc | Vespa Super |
| VLA1 | 145.45 cc | Vespa GL |
| VLB1 | 145 45 cc | Vespa Sprint |
| VSD1 | 180.69 cc | Rally |
| VSE1 | 197.97 cc | Rally 200 Electronic |

*All rotary valve models, from 1959 onwards*

**ISBN 978 0 85696 126 7**

Printed in Malaysia   *(126–4W7)*

ABCDE
FGH

3

**Haynes Publishing Group**
Sparkford Nr Yeovil
Somerset BA22 7JJ England

**Haynes Publications, Inc**
859 Lawrence Drive
Newbury Park
California 91320 USA

# About this manual

The author of this manual has the conviction that the only way in which a meaningful and easy to follow text can be written is first to do the work himself, under conditions similar to those found in the average household. As a result, the hands seen in the photographs are those of the author. Even the machines are not new; examples that have covered a considerable mileage were selected, so that the conditions encountered would be typical of those found by the average owner/rider. Unless specially mentioned and therefore considered essential, Vespa service tools have not been used. There is invariably alternative means of loosening or slackening some vital component, when service tools are not available but risk of damage is to be avoided at all costs.

Each of the six Chapters is divided into numbered Sections. Within the Sections are numbered paragraphs. Cross-reference throughout this manual is quite straightforward and logical. When reference is made, 'See Section 6.10' — it means Section 6, paragraph 10 in the same Chapter. If another Chapter were meant it would say 'See Chapter 2, Section 6.10'.

All photographs are captioned with a Section/paragraph number to which they refer, and are always relevant to the Chapter text adjacent.

Figure numbers (usually line illustrations) appear in numerical order, within a given Chapter. 'Fig 1.1' therefore refers to the first figure in Chapter 1.

Left hand and right hand descriptions of the machines and their components refer to the left and right of a given machine when normally seated facing the front wheel.

Motorcycle manufacturers continually make changes to specifications and recommendations, and these, when notified, are incorporated into our manuals at the earliest opportunity.

Whilst every care is taken to ensure that the information in this manual is correct no liability can be accepted by the authors or publishers for loss, damage or injury caused by any errors in, or omissions from, the information given.

# Acknowledgements

Our grateful thanks are due to both Eric Brockway and Roy Stone of Douglas (Sales and Service) Limited for the technical assistance given so freely whilst this manual was being prepared. We also wish to acknowledge the permission given by the Company to reproduce many of their drawings. Rod Grainger kindly provided the 150 cc Sportique model used for the photographic sequences and N. G. Preston Motorcycles of Yeovil supplied the necessary spare parts needed during the rebuild. Brian Horsfall assisted with the dismantling and re-building sequences, devising the many ingenious methods adopted for overcoming the lack of service tools. Les Brazier arranged and took the photographs. Tim Parker edited the text.

The cover photograph was arranged through the courtesy of Douglas (Sales and Service) Ltd of Bristol.

# Modifications to the Vespa scooter range

The basic design of the Vespa scooter has remained unchanged since its successful introduction to the UK during 1951. Engines modified to incorporate a rotary valve became available during 1959 and an alternative method of front suspension has been used on the smaller capacity models. In some of these latter models the clutch has been mounted on the end of the gearbox layshaft and not on the crankshaft as was the original practice. Other changes were of detail improvement or styling nature.

# Contents

The Vespa 150 Sportique Scooter

# Introduction to the Vespa scooter

It is a popular misconception that the scooter type of machine originated from Italy, just after World War II. Whilst this was undoubtedly when the present scooter trend commenced, it was by no means the first. The British motorcycle industry had experimented with scooter-type machines some thirty years earlier, soon after the 1914-18 war, having seen the potential for a machine that could be used as the stepping stone to the ownership of a fully-fledged motorcycle, or alternatively, as a particularly convenient method of covering short distances at low cost. There was even the possibility that many who were prejudiced against the conventional motorcycle would see the scooter as a more refined means of transport, without the need to wear special clothing or have a good mechanical knowledge. Sadly, this first bid to capture a new market failed for two very good reasons. Most of the designs had no provision for seating, so that the rider was forced to stand. Furthermore, the small capacity engine was often located where effective air cooling was not possible. This, and the fact that a gearbox was considered unnecessary, led to a standard of performance that was disappointing in the extreme. Those who had taken the plunge and bought scooters soon became disillusioned and within a very short period of time, the scooter became an historic relic.

After the second World War, news about the much more sophisticated Italian designs began to filter through, for some of the machines were on test as early as 1945. The publication of road tests in Britain's two motor cycle weeklies heightened interest, for it was apparent that a 125 cc fan-cooled two-stroke engine was capable of providing a surprisingly high standard of performance. Petrol economy was good, to match, and there was even a reasonable amount of in-built weather protection, together with provision for carrying a small amount of luggage.

Douglas (Sales and Service) Limited were quick to realise the potential of these new designs and after importing one of the Vespa models made by Piaggio and C. s.p.a. of Genoa, they announced plans to manufacture both the Vespa scooter and the Ape three-wheeled goods vehicle, under licence, at their Bristol factory. The announcement was made on the day preceding the opening day of the 1949 annual Motor Cycle Show at Earls Court. When the Show opened, the Vespa and Ape models on the Douglas stand carried the Douglas motif.

Production of the Douglas Vespa commenced at Hanham Road, Bristol, on March 15th 1951 and continued until 1964, when manufacture of the British-made models finally ceased. Thereafter, only the importation of Italian-made models continued; they were imported in knocked down form and reassembled for sale on the British market. The association between Piaggio and C that commenced during 1949 has remained substantially unchanged. Today, Douglas (Sales and Service) Limited operates from new factory premises in Bristol, on the Fishponds Trading Estate, and has difficulty in meeting the ever-continuing demand for Vespa scooters.

# Ordering spare parts

When ordering spare parts for any of the Vespa models, it is advisable to deal direct with an official Vespa agent who will be able to supply many items ex-stock and the remainder at very short notice. Always quote the engine and chassis numbers in full when ordering spares, particularly if the parts required are for any of the older models. The chassis number will be found on the right hand rear end of the chassis, underneath the detachable sidecover. It is stamped on a plate rivetted to the chassis panelling. The engine number is stamped close to the front engine mounting, in the vicinity of the exhaust pipe. Take note of any prefixes or suffixes to the numbers; they are just as important, for identification purposes.

Use only parts of genuine Vespa manufacture. Pattern parts are available, often at lower cost, but in many instances they will have an adverse effect on performance and/or reliability. Some complete units such as crankshaft assemblies, cylinder barrels complete with pistons, brake shoes, steering columns, chassis damper units and speedometer heads are available on a 'service exchange' basis from Douglas (Sales and Service) Limited. They afford an economic method of repair, provided the parts handed in can be reclaimed. Retain any broken or worn parts as a pattern for identification purposes, a problem that becomes more acute as the machine grows older. In an extreme case, it may be possible to reclaim the broken or worn part, or to use it as a pattern for making a replacement. Many older machines can be kept on the road in this manner, long after a manufacturer's spares have ceased to be available. Fortunately, there is no undue concern about spares for the older models at this time.

Some of the more expendable parts such as spark plugs, bulbs, tyres, oils and greases etc., can be obtained from accessory shops and motor factors, who have convenient opening hours, charge lower prices and can often be found not far from home. It is also possible to obtain parts on a Mail Order basis from a number of specialists who advertise regularly in the motor cycle magazines.

Frame number location

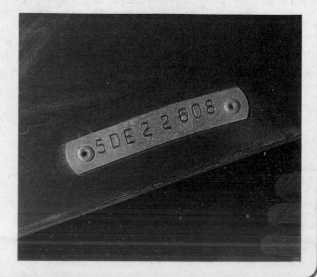

Engine number location

# Routine maintenance

Periodic routine maintenance is a continuous process that commences immediately the machine is used. It must be carried out at specified mileage recordings or on a calendar basis if the machine is not used frequently, whichever falls soonest. Maintenance should be regarded as an insurance policy, to help keep the machine in the peak of condition and to ensure long, trouble-free service. It has the additional benefit of giving early warning of any faults that may develop and will act as a regular safety check, to the obvious advantage of both rider and machine alike.

The various maintenance tasks are described under their respective mileage and calendar headings. Accompanying diagrams are provided, where necessary. It should be remembered that the interval between the various maintenance tasks serves only as a guide. As the machine gets older or is used under particularly adverse conditions, it would be advisable to reduce the period between each check.

Some of the tasks are described in detail, where they are not mentioned fully as a routine maintenance item in the text. If a specific item is mentioned but not described in detail, it will be covered fully in the appropriate Chapter. No special tools are required for the normal routine maintenance tasks. The tools contained in the kit supplied with every new machine will prove adequate for each task or if they are not available, the tools found in the average household.

## Three weekly or every 600 miles (1000 km)

With the engine warm, remove the crankcase drain plug and drain off all excess oil. On new machines or where the engine has been substantially rebuilt, introduce some clean oil through the drain plug orifice with a pressure oil can and permit this to drain off too. Replace the drain plug when no more oil is released, making sure the sealing washer is in good condition. On a new machine, the gearbox should be drained and refilled at the same time.

Check all nuts and bolts for tightness. Make sure the electrolyte level of the battery is not low. If it has fallen below the minimum level, top up with distilled water.

## Two monthly or every 2500 miles (4000 km)

With the engine warm, drain off the gearbox oil and refill with fresh oil of the correct viscosity. Replace the filler plug only after all excess oil has drained off.

Remove the air filter and clean it with petrol. Dry with compressed air. Grease the gear selectors, front wheel hub, brake controls and the felt that lubricates the cam of the flywheel magneto. Clean the spark plug electrodes with a wire brush and re-adjust the gap of 0.025 inch (0.6 mm).

Decarbonise the piston, cylinder head, cylinder ports and the external cylinder surfaces. Remove the silencer and exhaust pipe so that it can be cleaned out whilst it is heated, using a piece of bent wire as a probe. Point the exhaust pipe downward during this operation.

Lubricate the speedometer drive cable and the pinion. Make sure there is no lubricant on the last six inches of the cable, where it enters the speedometer head.

## Four monthly or every 5000 miles (8000 km)

Clean the contact breaker points and reset the gap to within the limits of 0.011 - 0.019 inch (0.3 - 0.5 mm). Recheck the accuracy of the ignition timing.

It should be noted that even when the two monthly or four monthly maintenance tasks have to be undertaken, the three weekly items must also be completed. There is no stage at any point during the life of the machine when a routine maintenance task can be ignored.

No special mention has been made relating to the lighting equipment, horn and speedometer, all of which must be in good working order if the statutory requirements of the UK are to be met. Regulations also apply to the minimum depth of tyre tread and the overall condition of the tyres. It is assumed every rider will keep a watchful eye on these additional points, especially since they have a direct bearing on rider safety. Regular safety checks should never be neglected.

# Routine maintenance and capacities data

| | |
|---|---|
| Engine | Self-mixing two-stroke oil. 2% petrol/oil mixture |
| Gearbox | SAE 30  Fill to level of filler plug |
| Contact breaker gap | 0.011 - 0.019 inch (0.3 - 0.5 mm) |
| Spark plug gap | 0.025 inch (0.6 mm) |
| Tyre pressures | 17 psi front  23 psi rear  (solo) - 3.00 inch tyres |
| | 16 psi front  20 psi rear  (solo) - 3.50 inch tyres |
| Fuel tank capacity | 1.15 - 1.7 Imp gallons (depending on model) |

# Recommended lubricants

| | |
|---|---|
| ENGINE  ...    ...    ...    ...    ...    ... | Castrol TT Two-Stroke Oil |
| GEARBOX    ...    ...    ...    ...    ... | Castrol GTX |
| FRONT SUSPENSION | |
| FELT PAD ON FLYWHEEL CAM | |
| JOINTS ON BRAKE CONTROLS | |
| SPEEDOMETER CABLE    ...    ... | Castrol LM Grease |
| CONTROL CABLES | |
| GEAR SELECTOR MECHANISM | |

# Chapter 1 Engine, clutch and gearbox

## Contents

## Specifications

| Model | 312L2 (Sportique) | 232L2 | V9A1 (Vespa 90) | VLA1 (Vespa GT) | VLB1 (Sprint) | V9SS1 (Vespa 90 Super Sport) |
|---|---|---|---|---|---|---|
| **Engine** | | | | | | |
| Cubic capacity ... ... ... ... ... ... ... ... | 145.45 | 123.4 | 88.5 | 145.45 | 145.45 | 88.5 |
| Cylinder bore (mm) ... ... ... ... ... ... ... | 57 | 47 | 47 | 57 | 57 | 47 |
| Piston stroke (mm) ... ... ... ... ... ... ... | 57 | 51 | 51 | 57 | 57 | 51 |
| Compression ratio ... ... ... ... ... ... ... | 6.5 : 1 | 7.2 : 1 | 7.2 : 1 | 7.2 : 1 | 7.5 : 1 | 8.7 : 1 |
| Bhp at rpm ... ... ... ... ... ... ... ... | 5.5 @ 5000 | — | 3.1 @ 5200 | — | 5.9 @ 5200 | 5.1 @ 5750 |
| Lubrication (Petroil) ... ... ... ... ... ... | 2% | 2% | 2% | 2% | 2% | 2% |
| **Piston** | | | | | | |
| Type ... ... ... ... ... ... ... ... ... | Deflector | Deflector | Domed top | Domed top | Domed top | Domed top |
| Oversizes available ... ... ... ... ... ... | | | + 0.2 mm, + 0.4 mm and + 0.6 mm | | | |
| Cylinder wear limit ... ... ... ... ... ... | +0.20 mm | +0.20 mm | +0.15 mm | +0.20 mm | +0.20 mm | +0.15 mm |
| **Piston rings** | | | | | | |
| Number ... ... ... ... ... ... ... ... | Two, with pegged ends — all models | | | | | |
| End gap ... ... ... ... ... ... ... ... | 0.2 - 0.35 mm (0.008 - 0.014 in) — all models | | | | | |
| Wear limit ... ... ... ... ... ... ... ... | 2.0 mm (0.080 in) — all models | | | | | |
| **Big end** | | | | | | |
| Maximum side float ... ... ... ... ... ... | 0.7 mm (0.28 in) — all models | | | | | |
| | 312L2 (Sportique) | 232L2 | V9A1 (Vespa 90) | VLA1 (Vespa GT) | VLB1 (Sprint) | V9SS1 (Vespa 90 Super Sport) |
| **Gear ratios** | | | | | | |
| Bottom gear ... ... ... ... ... ... ... ... | 13.35 : 1 | 12.2 : 1 | 17.18 : 1 | 14.46 : 1 | 14.46 : 1 | 14.74 : 1 |
| Second gear ... ... ... ... ... ... ... ... | 9.32 : 1 | 7.6 : 1 | 9.66 : 1 | 10.28 : 1 | 10.28 : 1 | 9.80 : 1 |
| Third gear ... ... ... ... ... ... ... ... | 6.64 : 1 | 4.85 : 1 | 6.12 : 1 | 7.46 : 1 | 7.36 : 1 | 7.06 : 1 |
| Fourth gear ... ... ... ... ... ... ... ... | 4.73 : 1 | — | — | 15.48 : 1 | 5.36 : 1 | 5.31 : 1 |

**Clutch**

Number of plates ... ... ... ... ... ... ... ... ...  ...................................... 3 friction, 2 plain .......................................

**Clutch springs**

| Number ... ... ... ... ... ... ... ... ... ... ... | 6 | 6 | 1 | 6 | 6 | 1 |
|---|---|---|---|---|---|---|

| Model | VBC1 (Vespa Super) | VMA1 (Vespa 125) | VMA2 (Primavera) | VSD1 (Rally) | V9SS2 (Racer) | VSE1 (Rally 200 Electronic) |
|---|---|---|---|---|---|---|
| **Engine** | | | | | | |
| Cubic capacity ... ... ... ... ... ... | 145.45 | 121.1 | 121.1 | 180.69 | 88.5 | 197.97 |
| Cylinder bore (mm) ... ... ... ... ... | 57 | 55 | 55 | 63.5 | 47 | 57 |
| Piston stroke (mm) ... ... ... ... ... | 57 | 51 | 51 | 57 | 51 | 66.5 |
| Compression ratio ... ... ... ... ... | 7.4 : 1 | 7.2 : 1 | 8.2 : 1 | 8 : 1 | — | 8.2 : 1 |
| Bhp at rpm ... ... ... ... ... ... | 5.7 @ 5200 | 4.3 @ 4750 | 4.5 @ 5700 | 8.7 @ 5750 | 5.1 @ 5750 | 9.8 @ 5700 |
| Lubrication (Petroil) ... ... ... ... | 2% | 2% | 2% | 2% | 2% | 2% |
| **Piston** | | | | | | |
| Type ... ... ... ... ... | Domed top | Domed top | Domed top | Domed top | Domed top | Domed top |
| Oversizes available ... ... ... ... | ............................ + 0.2 mm, + 0.4 mm and + 0.6 mm ............................. | | | | | |
| Cylinder wear limit | +0.20 mm | +0.15 mm | +0.15 mm | +0.25 mm | +0.15 mm | +0.25 mm |
| **Piston rings** | | | | | | |
| Number ... ... ... ... ... ... | Two, with pegged ends — all models | | | | | |
| End gap ... ... ... ... ... ... | 0.2 - 0.35 mm (0.008 - 0.014 in) — all models | | | | | |
| Wear limit ... ... ... ... ... ... | 2.0 mm (0.080 in) — all models | | | | | |
| **Big end** | | | | | | |
| Maximum side float ... ... ... ... | 0.7 mm (0.28 in) — all models | | | | | |

| | VBC1 (Vespa Super) | VMA1 (Vespa 125) | VMA2 (Primavera) | VSD1 (Rally) | V9SS2 (Racer) | VSE1 (Rally 200 Electronic) |
|---|---|---|---|---|---|---|
| **Gear ratios** | | | | | | |
| Bottom gear ... ... ... ... ... ... | 13.35 : 1 | 14.74 : 1 | 14.74 : 1 | 14.47 : 1 | 14.74 : 1 | 13.42 : 1 |
| Second gear ... ... ... ... ... ... | 9.32 : 1 | 9.80 : 1 | 9.80 : 1 | 9.84 : 1 | 9.80 : 1 | 9.13 : 1 |
| Third gear ... ... ... ... ... ... | 6.64 : 1 | 7.06 : 1 | 7.06 : 1 | 6.81 : 1 | 7.06 : 1 | 6.32 : 1 |
| Fourth gear ... ... ... ... ... ... | 4.73 : 1 | 5.31 : 1 | 5.31 : 1 | 5.08 : 1 | 5.31 : 1 | 4.71 : 1 |

**Clutch**

Number of plates ... ... ... ... ... ... ... ... ...  ...................................... 3 friction, 2 plain .......................................

**Clutch springs**

| Number ... ... ... ... ... ... ... ... ... | 6 | 1 | 1 | 6 | 1 | 6 |
|---|---|---|---|---|---|---|

## 1 General description

The engine unit fitted to the range of Vespa scooters covered by this manual is single cylinder two-stroke. The engine is built in unit with the clutch and gearbox and forms an integral part of the rear suspension. Final drive is taken direct from the gearbox layshaft and in consequence the rear wheel is also part of the engine/gear unit.

All castings are aluminium alloy, with the exception of the cylinder barrel, which is cast iron. A conventional, five port arrangement is used, with two inlet ports, two transfer ports and a single exhaust port.

Induction is aided by a rotary valve arrangement, using a cutaway in the flywheel assembly to unmask the induction passageway in the crankcase wall. It is thus possible to control the induction sequence more finely and place less reliance on the deflector on top of the piston crown, if the model is fitted with a piston of this type.

All models are fitted with a flywheel generator which supplies the ignition and lighting systems. It is mounted on the right hand side of the engine. A fan is attached to the generator rotor to air cool the engine, which is otherwise enclosed by the chassis of the machine. The exhaust system is short and uses a very efficient flat box-like silencer. All models are fitted with a kickstarter pedal which operates from the right hand side of the engine unit.

Lubrication of the engine is effected by the petroil system, in which oil in a prescribed quantity is pre-mixed with the petrol contained in the petrol tank. This method can be used very successfully in a two-stroke engine because the incoming mixture is first delivered to the crankcase, where it is compressed before it enters the transfer ports and passes into the cylinder. It is therefore easy to ensure that all the major engine components are adequately lubricated. The gearbox has its own separate oil content.

All engines are constructed in unit with the gearbox and form an integral part of the rear suspension system. This means that when the engine is dismantled completely, the clutch and gearbox are dismantled too and the whole swinging arm unit is detached from the chassis. This task is made easy by following a simple routine.

## 2 Operations with engine unit in chassis

1  It is not necessary to remove the engine unit from the chassis unless the crankshaft assembly and/or the gearbox bearings or pinions require attention. Most operations can be accomplished with the engine in place, such as:

a) Removal and replacement of the cylinder head
b) Removal and replacement of the cylinder barrel
c) Removal and replacement of the flywheel generator

2  Where several operations have to be undertaken simultaneously, it will probably be advantageous to remove the complete engine/gear unit from the chassis, an operation which should take approximately an hour. This will give the advantage of better access and more working space.

## 3 Operations with engine unit removed

1  Removal and replacement of main bearings.
2  Removal and replacement of crankcase oil seals.

3 Removal and replacement of the crankshaft assembly.
4 Removal and replacement of the gearbox internals and the gearbox bearings.
5 Removal of the clutch.

### 4 Method of engine unit removal

As mentioned previously, the engine and gearbox are built in unit and form an integral part of the swinging arm suspension. It is necessary to remove the complete swinging arm assembly and to separate the castings before access is available to either the engine or the gearbox. Separation cannot take place until the swinging arm assembly has been detached from the chassis and refitting cannot take place until the engine and gearbox are built up and the castings reassembled.

### 5 Removing the engine/gearbox unit

1 Place the machine on the centre stand so that it is standing evenly on level ground. Remove the two bolts which retain the exhaust system in position; one is located close to the exhaust port of the cylinder barrel and the other holds the silencer box to the underside of the chassis.
2 Place a receptacle under the gearbox and remove the hexagon headed drain plug which is now exposed. Allow the oil to drain off whilst attention is given to other items.
3 Withdraw the split pin from the castellated nut in the centre of the rear wheel, then unscrew the nut and remove the washer on which it seats. Raise the rear end of the machine by placing a wooden prop under the extreme end of the tail section, then pull the wheel off the splined centre shaft. It will come away complete with brake drum and driving flange and should be withdrawn at an angle after detachment, so that there is sufficient clearance for it to be freed from the chassis.
4 Disconnect the clutch and brake cables. The clutch cable is released by depressing the clutch operating arm so that the cable nipple can be lifted out of the end. The brake cable is released by slackening the split clamp on the end of the brake operating arm. Before either cable can be withdrawn completely, the adjuster must be unscrewed from the engine unit casting.
5 Take off the kickstarter, which is retained by a pinch bolt. Before it is pulled off the splined shaft, mark both the shaft and the kickstarter arm, so that it is eventually replaced in the same position. Remove the right hand 'blister' cover from the chassis which is held in position by a single screw.
6 Access is now available to the gearchange assembly. This is removed from the gearbox by detaching the two nuts and washers and drawing it off along the studs. It will be necessary to rotate the assembly when it is clear of the studs and possibly to operate the handlebar gear control as the assembly is withdrawn, in order to help disengage the selector skid from the control rod which protrudes from the gearbox. When free, the assembly can be lifted away and left suspended from the cables.
7 On some models this technique does not apply and it is necessary to detach both control cables as the alternative. The cables are attached by means of clamping nipples which release the cables when they are slackened off.
8 Remove the lid from the air cleaner and carburettor compartment and detach the throttle and choke cables. They are held by wire clips. Disconnect the fuel pipe from the carburettor at the union joint and pull off the convoluted hose which joins the air cleaner box to the air intake on the chassis.
9 Lower the hinged flap on the left hand 'blister' cover of the chassis and detach the positive lead from the battery to isolate the electrical system. Then, reverting to the right hand side of the machine, disconnect the lead wires from the coil and those to the junction box mounted on the flywheel generator casing.
10 The engine unit complete with the swinging arm assembly can now be removed from the chassis. Remove the bolt through the lower end of the rear suspension unit and drop the rear end of the assembly downward. Then place a support under the

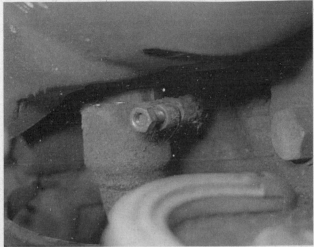
5.1a Exhaust system clamps to exhaust stub of cylinder barrel

5.1b Expansion box bolts to underside of chassis

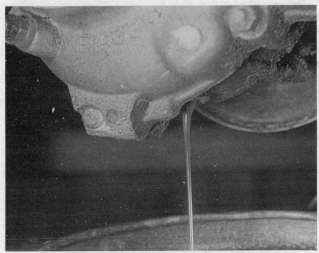
5.2 Drain gearbox oil before removing engine unit

gearbox so that the assembly is near horizontal. Detach the forward mounted bolt which is located immediately in fron tof the 'blister' covers on each side of the chassis. It is advisable to have someone support the engine unit assembly when this latter bolt is withdrawn. If the engine unit is kept horizontal, it can be drawn away from the chassis as a complete assembly, if necessary by raising the chassis a little to give extra clearance. Note that the wooden prop supporting the chassis must remain in position throughout this operation.

11 Transfer the engine unit assembly to a bench or some other convenient working area, where it can be cleaned thoroughly before further dismantling takes place.

5.3a Remove split pin from castellated nut

## 6 Dismantling the engine and gearbox - general

1 Before commencing work on the engine unit, the external surfaces must be cleaned thoroughly. A scooter engine is especially vulnerable to an accumulation of road grit and other foreign matter by virtue of its exposed position. Sooner or later, some of this dirt will find its way into the dismantled engine if this simple precaution is not observed.

2 One of the proprietary cleaning compounds such as Gunk or Jizer can be used to good effect, especially if the compound is first allowed to penetrate the film of grease and oil before it is washed away. Before applying the compound, it is best to remove all surplus dirt by scraping with a knife or similar object, so that the cleaning compound will have a better chance of penetration. When washing down, make sure that water cannot enter the carburettor or the electrical system, particularly now that these parts are more exposed.

3 Collect together a good set of tools including a set of metric spanners and a screwdriver with a blade of the correct size. Work on a clean surface and have a supply of clean, lint-free rag available.

4 Never use force to remove any stubborn part unless specific mention is made of this requirement in the text. There is invariably good reason why a part is difficult to remove, often because the dismantling procedure has been tackled in the wrong sequence.

5 Dismantling will be made easier if a simple engine stand is constructed or is purchased from the concessionaires. This will enable the complete unit to be clamped rigidly to the workbench, leaving both hands free for the dismantling operation.

5.3b Wheel will pull off splined shaft

## 7 Dismantling the engine and gearbox - removing the carburettor

1 It is necessary first to remove the air cleaner, which is attached to the carburettor by two screws which pass through the air cleaner body. There is no necessity to disturb the screw with the milled edge which stands proud of the air cleaner, since this is the throttle stop adjustment screw for the carburettor. It will pass through the air filter body and if removed or changed in setting, it will later be necessary to reset the carburettor.

2 The carburettor is retained by two shouldered nuts which pass through the carburettor body and locate with studs in the base of the air cleaner compartment. Withdraw both nuts and lift the carburettor away as a complete unit.

3 Before the base of the air cleaner box can be removed, it is necessary to remove the single retaining screw which is hidden below the gasket between the base of the carburettor and its mounting flange.

## 8 Dismantling the engine and gearbox - removing the flywheel generator

1 Remove the cowling which surrounds the cylinder head and barrel. It is retained by a single bolt which threads into an extension of one of the cylinder head nuts.

2 Detach the outer casing of the flywheel generator, which is retained by four screws around the periphery. Then remove the

5.6 Withdraw gear selectors as a complete unit

5.8a Remove air cleaner lid for access to carburettor

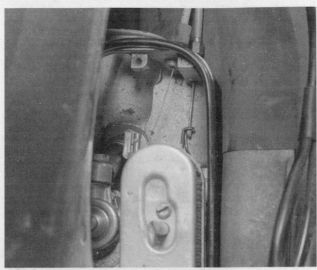

5.8b Carburettor cables are retained by clips

5.9 Isolate battery by removing negative lead

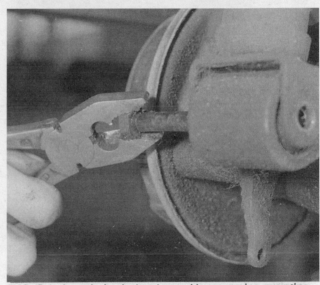

5.10a Bolt through shock absorber end is rear engine mounting point

5.10b Allow engine unit to drop before removing front bolt

5.10c Front bolt passes through chassis

7.1 Lift off air cleaner after releasing two screws

7.2 Carburettor is retained by two sleeve nuts

7.3a Carburettor gasket masks countersunk retaining screw

7.3b When retaining screw is removed, base of air cleaner compartment will lift off

8.1 Engine cowl is retained by a single bolt

8.2a Take off fan casing for access to rotor

8.2b Four bolts retain fan to flywheel generator

8.3a Flywheel rotor has self-extracting nut

8.3b Contact breaker assembly is behind flywheel rotor

8.5 Mark position of stator plate before it is removed

fan which is bolted to the centre of the flywheel by four bolts, each with a shakeproof washer. The flywheel itself is now exposed.
3   The flywheel has a self-extracting nut. To release the flywheel from its taper, slacken the nut until it comes into contact with the retaining circlip. Strike the nut a sharp blow with a copper or hide-faced hammer in conjunction with a soft alloy drift. Then continue slackening the nut. It may be necessary to repeat the procedure several times before the flywheel is free. On no account drive the nut so that it forces the circlip out of its locating groove. Immediately the flywheel is free, bridge the magnets with soft iron to prevent loss of magnetism.
4   Remove the HT coil (if fitted) which is clamped to the rear half of the flywheel generator housing. Disconnect the remainder of the electrical leads from the junction box. Make a scribe mark across the stator plate and the crankcase to ensure that the stator plate is aligned correctly during reassembly.
5   The stator plate is retained by three screws around the periphery. Remove all three screws, then pull the electrical leads through the hole in the rear of the generator housing. The stator plate assembly can now be withdrawn and should be placed within the flywheel, where it will fulfill the role previously accomplished by the soft iron. Remove the Woodruff key from the crankshaft and place it in a container for safe keeping.

## 9  Dismantling the engine and gearbox - removing the brake shoes

1   It is necessary to remove the brake shoes and the brake plate

to provide better access to the clutch housing. Remove the circlip from the brake shoe pivot, then lift off the brake shoes after slipping the ends off the brake operating arm.
2   The brake plate will lift off when the brake shoes have been removed. It is a light push fit over the hub centre.

## 10  Dismantling the engine and gearbox - removing the clutch

1   Detach the domed clutch cover, which is retained by three screws. Do not lose the plunger which will be displaced from the clutch cover as the latter is lifted away.
2   The centre of the clutch thrust plate is held in position by a wire clip. Detach the clip and lift out the circular insert which will expose the castellated nut which locks the clutch on the end of the crankshaft.
3   Bend back the tab washer which engages with the castellated nut and use either Vespa service tools T0029551 (wrench) and T0019354 (spanner) or a home-made peg spanner as shown in the accompanying illustration. The nut should not prove to be exceptionally tight. When the nut has been withdrawn from the clutch centre, the complete clutch unit can be lifted off the end of the crankshaft using two screwdrivers as levers. It should not be necessary to use excessive force to break the parallel shouldered joint; beware of marking or indenting the clutch cover joint during this operation. The clutch will lift off complete with the helical driving pinion which is an integral part of the main body.

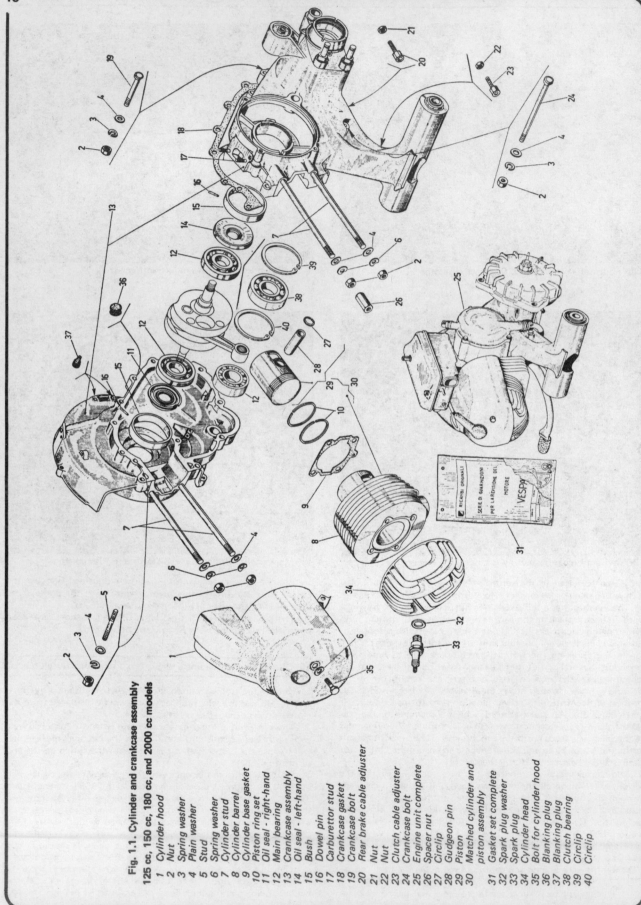

**Fig. 1.1. Cylinder and crankcase assembly**

**125 cc, 150 cc, 180 cc, and 2000 cc models**

1  Cylinder hood
2  Nut
3  Spring washer
4  Plain washer
5  Stud
6  Spring washer
7  Cylinder stud
8  Cylinder barrel
9  Cylinder base gasket
10  Piston ring set
11  Oil seal - right-hand
12  Main bearing
13  Crankcase assembly
14  Oil seal - left-hand
15  Bush
16  Dowel pin
17  Carburettor stud
18  Crankcase gasket
19  Crankcase bolt
20  Rear brake cable adjuster
21  Nut
22  Nut
23  Clutch cable adjuster
24  Crankcase bolt
25  Engine unit complete
26  Spacer nut
27  Circlip
28  Gudgeon pin
29  Piston
30  Matched cylinder and
    piston assembly
31  Gasket set complete
32  Spark plug washer
33  Spark plug
34  Cylinder head
35  Bolt for cylinder hood
36  Blanking plug
37  Blanking plug
38  Clutch bearing
39  Circlip
40  Circlip

Fig. 1.2. Cylinder and crankcase assembly
90 cc V9A1 and V9SS1 models and 125 cc VMA1 and VMA2 models only

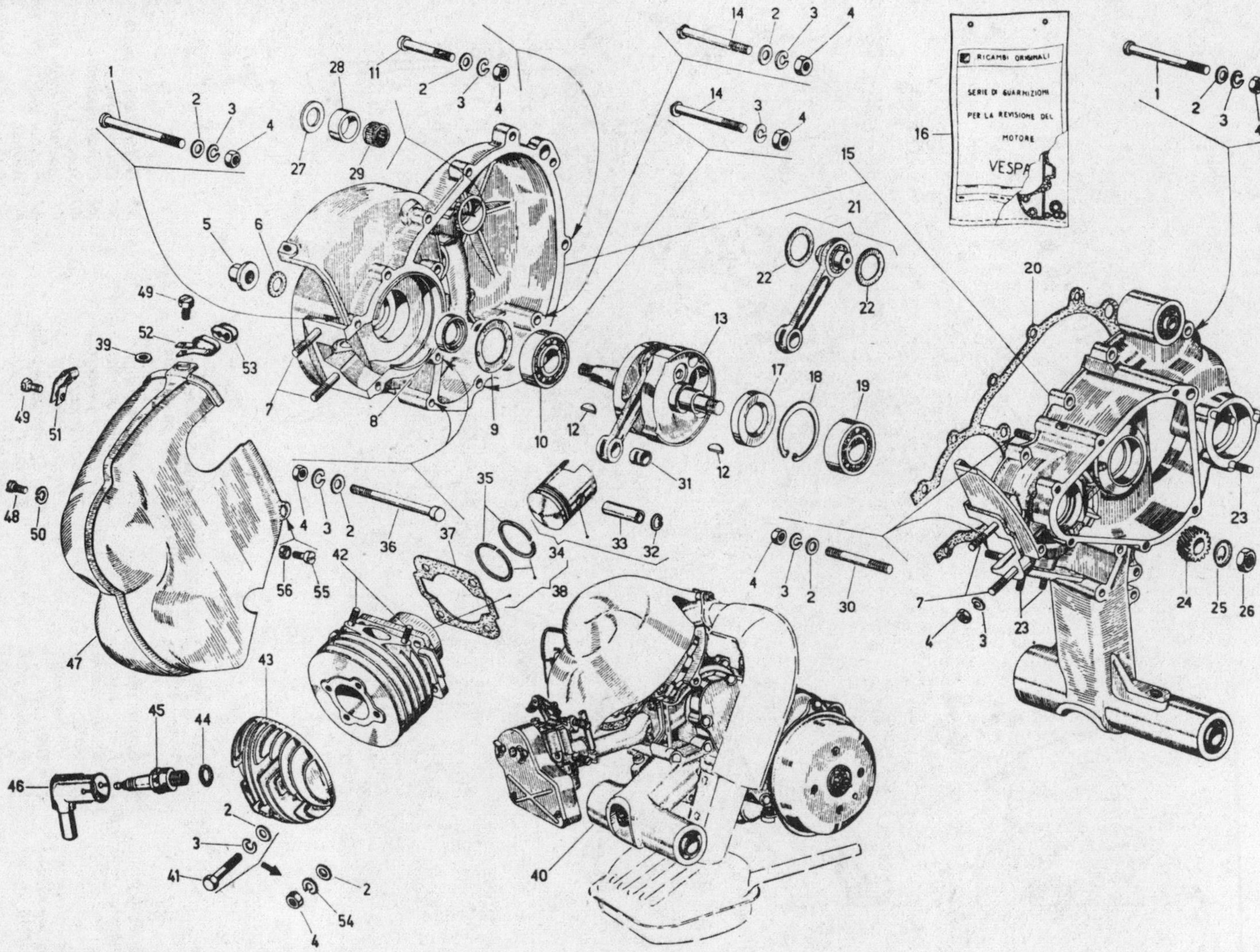

1  Bolt
2  Plain washer
3  Spring washer
4  Nut
5  Nut and collar
6  Star washer
7  Stud
8  Oil seal, right-hand side
9  Spacer washer
10  Main bearing, right-hand side
11  Bolt, short
12  Key
13  Crankshaft
14  Bolt
15  Crankcase
16  Gasket set complete
17  Oil seal, left-hand side
18  Circlip
19  Main bearing, left-hand side
20  Gasket
21  Connecting rod and crankpin
22  Spacer washer
23  Spacer washer
24  Clutch drive pinion
25  Stop washer
26  Nut
27  Spacer washer
28  Bearing track
29  Caged needle roller bearing
30  Stud
31  Circlip
32  Gudgeon pin
33  Piston and gudgeon pin
34  Piston ring set
35  Spacer washer
36  Bolt, long
37  Cylinder base gasket
38  Cylinder, piston and gudgeon pin
39  Cooling hood securing washer
40  Engine unit complete
41  Cylinder head retaining stud
42  Exhaust pipe securing stud
43  Cylinder head
44  Spark plug washer
45  Spark plug
46  Suppressor cap
47  Cooling hood
48  Cooling head securing screw
49  Cable securing screws
50  Spring washer
51  Cable securing strap
52  Strap
53  Buffer securing cables
54  Spring washer for cylinder head
55  Screw securing cooling hood
56  Spacer for cooling hood

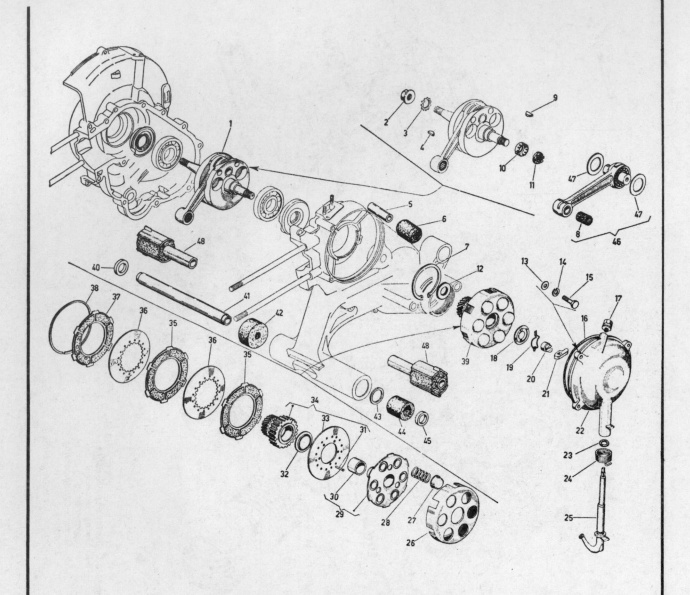

**Fig. 1.3. Crankshaft and clutch assemblies
123 cc, 150 cc, 180 cc and 200 cc models**

| | | | |
|---|---|---|---|
| 1 Crankshaft assembly | 12 Spacer washer | 24 Return spring | 37 Outer friction plate |
| 2 Flywheel generator securing nut | 13 Plain washer | 25 Clutch actuating arm | 38 Clutch circlip |
| 3 Star washer | 14 Spring washer | 26 Clutch pressure plate | 39 Clutch assembly |
| 4 Flywheel generator key | 15 Bolt | 27 Spring cup | 40 Spacer |
| 5 Engine mounting rubber | 16 'O' ring seal | 28 Clutch spring | 41 Mounting inner tube |
| 6 Engine mounting rubber | 17 Breather | 29 Clutch back plate | 42 Rubber buffer -right-hand |
| 7 Circlip | 18 Pressure plate insert | 30 Bush | 43 Washer |
| 8 Small end bearing | 19 Spring clip | 31 Rivet | 44 Rubber buffer left-hand |
| 9 Clutch key | 20 Actuating pad | 32 Spacer washer | 45 Spacer |
| 10 Tab washer | 21 Actuating bell crank | 33 Clutch drive plate | 46 Connecting rod and crankshaft |
| 11 Sleeve nut | 22 Clutch cover | 34 Clutch drive pinion | 47 Spacer washer |
| | 23 Washer | 35 Friction plate | 48 Silent Bloc buffer |
| | | 36 Plain plate | |

9.1a Wire clip retains fixed ends of brake shoes

9.1b Lift shoes from brake plate in manner shown

10.1 Domed clutch cover contains actuating mechanism

10.2 Remove wire clip for access to clutch centre nut

10.3a Centre nut will require special peg spanner

10.3b Use screwdrivers to lift clutch from shaft

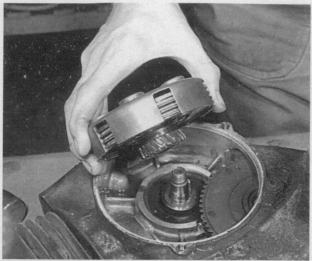

10.3c Clutch will withdraw as a complete unit

10.4 Don't lose thrust washer behind clutch assembly

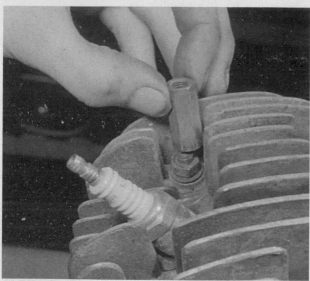

11.1 Remove sleeve nut from cylinder head first

11.2 Cylinder barrel will slide up studs

11.3a Discard circlips after removal

11.3b If gudgeon pin is tight, warm piston before removal

4   Remove the Woodruff key from the crankshaft and the thrust washer (where fitted).

5   The 90 cc and 125 cc models have the clutch fitted to the tapered end of the gearbox layshaft. Although the same basic dismantling procedure applies, it is necessary to use Vespa service tool T0029551 as the extractor, after removing the securing nut. The crankshaft pinion is removed by releasing the lockwasher and the retaining nut.

### 11   Dismantling the engine and gearbox - removing the cylinder head, barrel and piston

1   The cylinder head is retained by four nuts, one of which has an extension to provide the anchorage for the engine cowling. Remove this extension and each of the four nuts and the flat washer on which each seats. The cylinder head can then be removed. Note that no sealing gasket is employed at the cylinder head/barrel joint which is a ground, recessed fit.

2   Slide the cylinder barrel upward along the holding down studs, taking care to catch the piston as it emerges from the bore. If the engine is not to be stripped completely, it is advisable to pad the mouth of the crankcase with clean rag before the piston is disengaged from the bore. This will prevent particles of piston ring from dropping into the crankcase in the event of ring breakages.

3   To remove the piston, first detach one of the gudgeon pin circlips, then tap the gudgeon pin out of position whilst the piston is steadied. If the pin is a tight fit, a rag soaked in warm water placed on the piston crown will help expand the piston and ease the grip on the pin. When the piston is removed mark the inside of the skirt to show which way the piston faces. This will obviate the risk of the piston being replaced in the reverse position, an important consideration when the piston crown embodies a deflector. A reversed piston will cause a surprising loss of power, the cause of which will prove very difficult to detect. If the piston does not have a deflector, the crown will be stamped with an arrow which **must** face the exhaust. Reversal will bring the ends of the piston rings into contact with the ports and cause an engine seizure or ring breakage.

4   Remove the caged needle roller bearing from the small end of the connecting rod to prevent it falling out and getting lost.

### 12   Dismantling the engine and gearbox - separating the crankcases

1   If the gear selector box is still in position, it must be detached before the crankcases can be separated.

2   Remove the twelve bolts which hold together the two crankcase halves. Before separation can occur, it will be necessary to heat the right hand crankcase in the vicinity of the main bearing housing, using a gas blowlamp or some similar means of providing local heat. Heating is necessary because the main bearings are a shrink fit in their housings. Beware of overheating, or there is risk of the crankcase seals suffering irreparable damage.

3   When the bearing housing has expanded sufficiently to release the bearing, the crankcases can be separated with ease. It is essential, however, to ensure that the kickstarter quadrant is depressed, otherwise the quadrant will not have sufficient clearance and will impede separation. A mole wrench can be used to good effect, as shown in the accompanying illustration, provided there is no risk of damaging the splined shaft.

4   Never use a screwdriver or other sharp instrument to prise the crankcases apart. If they do not part easily, either the bearing is still tight in its housing or the kickstarter ratchet has not been depressed sufficiently. If either of the crankcase jointing faces is marked, an air and/or oil leak will occur after reassembly which may have a serious effect on engine performance through loss of crankcase compression or an air leak that dilutes the incoming mixture.

11.3c Mark piston inside skirt to ensure replacement in correct position

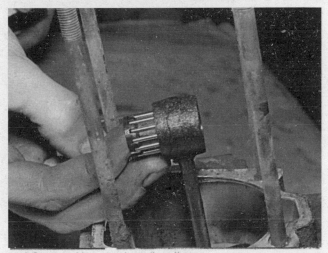

11.4 Small end has caged needle rollers

12.3a Depress kickstarter quadrant whilst crankcase is separated

### 13  Dismantling the engine and gearbox - removing the gearbox components

1   Lift off the small twelve tooth pinion from the end of the gearbox layshaft.

2   Take off the circlip and tongued washer from the centre of the gearbox mainshaft outer pinion, then withdraw the pinion and all the other pinions behind it, making note of their order of assembly.

3   To remove the cruciform gear selector, clamp the mainshaft in a vice fitted with soft clamps, straighten the selector rod tab washer and turn the rod in a clockwise direction by placing a spanner across the flats provided. Note this is a **left hand thread**. When unscrewed, the selector rod can be lifted out, followed by the guide bush. If the internal cruciform is permitted to drop into the horizontal position, it can be withdrawn through one of the slots in the mainshaft.

4   To release the mainshaft, invert the crankcase and remove the wire clip from the locking ring around the end of the shaft which engages with the centre of the rear wheel. Unscrew the locking ring (noting that it has a left–hand    thread) then temporarily replace the wheel retaining nut in reverse, to protect the thread and end of the shaft whilst it is driven through the bearing centre from the outside.

5   The layshaft is built as a multiple gear assembly. To remove the assembly from the mainshaft, unscrew the layshaft retaining nut located on the outside of the crankcase and drive the shaft inward. Take care to catch the 23 uncaged needle rollers which will be displaced as the shaft clears the gear assembly. Note that the left hand side main bearing is contained within the centre of the large diameter helical drive gear. It is retained by a circlip and can be driven out of position when the circlip is removed.

### 14  Dismantling the engine and gearbox - removing the crankshaft assembly

1   To remove the crankshaft assembly, heat the left hand crankcase so that the bearing housing will expand and release the shrunk-in main bearing. The crankshaft assembly can then be pulled from the crankcase and put aside for further attention at a later stage.

2   Do not use force to extract the crankshaft. If the bearing is a tight fit, it may be necessary to give the end of the crankshaft a few light taps with a rawhide mallet to start it moving.

3   If the crankshaft assembly pulls out leaving the main bearing in position, the bearing can be removed afterwards. In this case, the circlip in front of the large diameter oil seal should be removed first, then the oil seal. After the crankcase has been warmed, the bearing can be driven out of its housing using a soft metal drift and a rawhide mallet.

4   The kickstarter quadrant assembly will knock out of the right hand crankcase with a few taps from a hide mallet.

### 15  Examination and renovation - general

1   Before examining the parts of the dismantled engine for wear, it is essential that they should first be cleaned thoroughly. Use a petrol/paraffin mix to remove all traces of old oil and sludge that may have accumulated within the engine.

2   Examine the various castings for cracks or other signs of damage, especially the crankcase castings. If a crack is discovered, it will require professional repair, or renewal.

3   Examine carefully each part to determine the extent of wear, checking with the tolerance figures listed in the Specifications Section of this Chapter. If there is any question of doubt, play safe and renew. Notes included in the following text will indicate what type of wear can be expected and whether the part concerned can be reclaimed.

4   Use a clean, lint-free rag for cleaning and drying the various components. This will obviate the risk of small particles obstructing the internal oilways, causing the lubrication system to fail.

5   Above all, work in clean, well-lit surroundings so that faults do not pass undetected. Failure to detect a fault or signs of

12.3b Bearings will probably remain on crankshaft

13.1 Lift off kickstarter pinion

13.2a Remove circlip from gearbox mainshaft

13.2b Tongued washer below circlip must also be removed

13.2c Withdraw all gearbox pinions, noting order of assembly

13.2d Note how pinion boss faces outwards

13.3a Selector rod holder has a left-hand thread

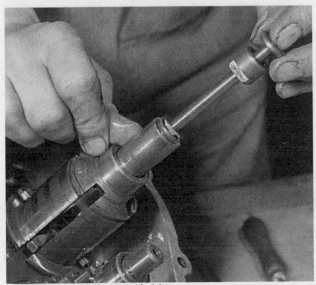
13.3b Lift out rod and rod holder ...

13.3c ... followed by guide bush

13.3d Cruciform will pass through slot in mainshaft

13.4a Wire clip locks bearing retainer

13.4b Unscrew bearing retainer for access to bearing

13.4c Nut on end of mainshaft prevents thread damage

10.6d Layshaft uses uncaged needle roller bearing at RH end

10.6b Layshaft is a built-up unit; cannot be dismantled

13.5c LH layshaft bearing is in centre of large drive pinion

14.1 Heat may be necessary to free bearings from crankcase

advanced wear may necessitate a further complete stripdown at a later date, due to the premature failure of the part concerned.

## 16 Main bearings and oil seals - examination

1  If the main bearings have come away with the crankshaft assembly, it is essential to use a bearing extractor to remove them. Although it may be possible to use a pair of screwdrivers to lever them off, there is grave risk of distorting the crankshaft assembly, which has a pressed-in crankpin. The crankshaft assembly needs a 0.002 inch clearance with the crankcase wall for the rotary valve to function efficiently. If the assembly is distorted, it will rub the crankcase and cause irreparable damage.
2  The right hand oil seal is still located in the right hand crankcase where it is a light drive fit. Take care not to damage the housing as the oil seal is driven out of position.
3  Wash the main bearings in a petrol/paraffin mix to remove all traces of oil. If there is any play, or if they do not revolve smoothly, both bearings should be renewed as a pair. It is false economy to renew only the one that seems to be defective because the other will fail soon after and necessitate a further complete engine strip.
4  Even if the oil seals appear to be in good condition, it is a wise precaution to renew them both. Failure of the oil seals is a common two-stroke malady because worn seals admit air to the crankcase which dilutes the incoming mixture before it is compressed or permit the escape of the charge whilst it is under compression. Crankcase air leaks are one of the most frequent causes of difficult starting and uneven running in any two-stroke engine.

## 17 Crankshaft assembly - examination and renovation

1  Wash the complete flywheel assembly with a petrol/paraffin mix to remove all surplus oil. Then hold the connecting rod at its highest point of travel (fully extended) and check whether there is any vertical play in the big end bearing by alternately pulling and pushing in the direction of travel. If the bearing is sound, there should be no play whatsoever.
2  Ignore any sideplay unless this appears to be excessive. A certain amount of play in this direction is necessary if the bottom end of the engine is to run freely.
3  Although it may be possible to run the engine for a further short period of service with a very small amount of play in the big end bearing, this course of action is not advisable. Apart from the danger of the connecting rod breaking if the amount of wear increases rapidly, a further complete engine strip will be necessary to effect the renewal. It is best to renew the big end bearing at this stage, if it is in any way suspect. Wear is denoted

by the characteristic 'knock' when the engine is running under load.
4  A replacement crankshaft assembly is available through the Vespa Service Exchange Scheme. A factory-reconditioned assembly will be supplied at an advantageous price against receipt of the worn but otherwise undamaged old assembly. This will obviate the delay if the old assembly has to be dismantled, a new big end fitted and then reassembled and aligned to a high standard of accuracy, using a lathe.

## 18 Cylinder barrel - examination and renovation

1  If the engine has seen a reasonable amount of service, there will probably be a lip at the uppermost end of the cylinder barrel, which marks the limit of travel of the top piston ring. The depth of the lip will give some indication of the amount of bore wear that has taken place, even though the amount of wear is not evenly distributed.
2  Remove the rings from the piston taking great care as they are brittle and very easily broken. There is more tendency for rings to gum in their grooves in a two-stroke engine. Insert one of the piston rings in the bore and press it downward with the piston so that it is square in the bore and is located about ¾ inch from the top. Measure the end gap with a feeler gauge. If it is more than 0.012 inch (0.3 mm) it is probable that the cylinder needs a rebore or renewal by service exchange. As in the case of the crankshaft assembly, the Vespa Service Exchange Scheme will provide a reconditioned cylinder and a matched piston complete with rings, on receipt of the old, worn originals.
3  Give the cylinder barrel a close visual inspection. If the surface of the bore is scored or grooved, indicative of an earlier seizure or a displaced circlip and gudgeon pin, a rebore or service exchange renewal is essential. Compression loss will have a very marked effect on performance.
4  Check that the outside of the cylinder barrel is clean and free from road dirt. Use a wire brush on the cooling fins if they are obstructed in any way. The engine will overheat badly if the cooling area is obstructed in any way. The application of matt cylinder black will help improve heat radiation.
5  Clean all carbon deposits from the exhaust ports and try and obtain a smooth finish in the ports without in any way enlarging them or altering their shape. The size and position of the ports predetermines the characteristics of the engine and unwarranted tampering can produce very adverse effects. An enlarged or re-profiled port does not necessarily guarantee an increase in performance.
6  Examine the working surface of the rings. If discoloured areas are evident, the rings should be renewed since the patches indicate the blow-by of gas. Check that there is not a build-up

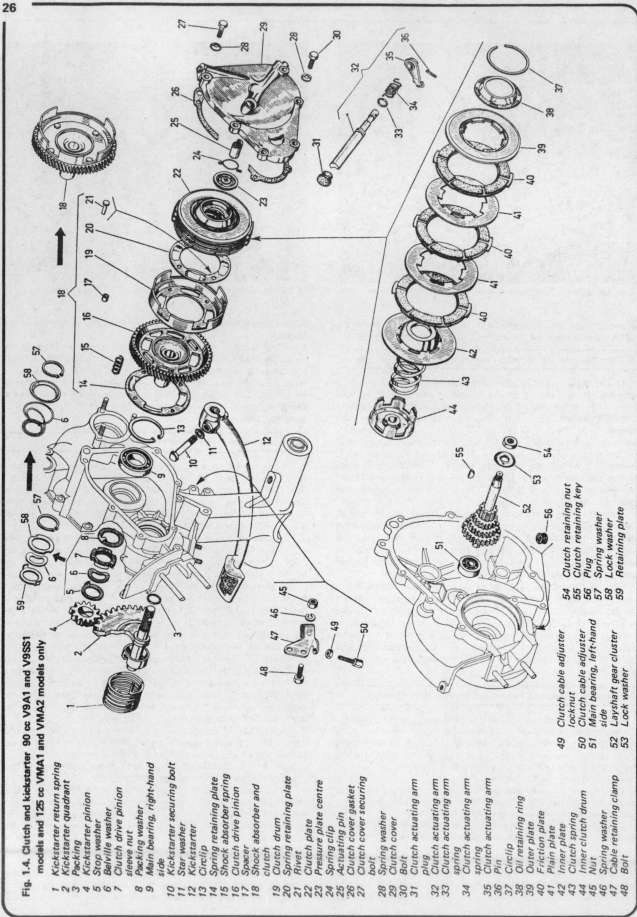

Fig. 1.4. Clutch and kickstarter 90 cc V9A1 and V9SS1 models and 125 cc VMA1 and VMA2 models only

1  Kickstarter return spring
2  Kickstarter quadrant
3  Packing
4  Kickstarter pinion
5  Stop washer
6  Belville washer
7  Clutch drive pinion sleeve nut
8  Packing washer
9  Main bearing, right-hand side
10  Kickstarter securing bolt
11  Star washer
12  Kickstarter
13  Circlip
14  Spring retaining plate
15  Shock absorber spring
16  Clutch drive pinion
17  Spacer
18  Shock absorber and clutch
19  Clutch drum
20  Spring retaining plate
21  Rivet
22  Clutch plate
23  Pressure plate centre
24  Spring clip
25  Actuating pin
26  Clutch cover gasket
27  Clutch cover securing bolt
28  Spring washer
29  Clutch cover
30  Bolt
31  Clutch actuating arm plug
32  Clutch actuating arm
33  Clutch actuating arm spring
34  Clutch actuating arm spring
35  Clutch actuating arm
36  Pin
37  Circlip
38  Oil retaining ring
39  Outer plate
40  Friction plate
41  Plain plate
42  Inner plate
43  Clutch spring
44  Inner clutch drum
45  Nut
46  Spring washer
47  Cable retaining clamp
48  Bolt
49  Clutch cable adjuster locknut
50  Clutch cable adjuster
51  Main bearing, left-hand side
52  Layshaft gear cluster
53  Lock washer
54  Clutch retaining nut
55  Clutch retaining key
56  Plug
57  Spring washer
58  Lock washer
59  Retaining plate

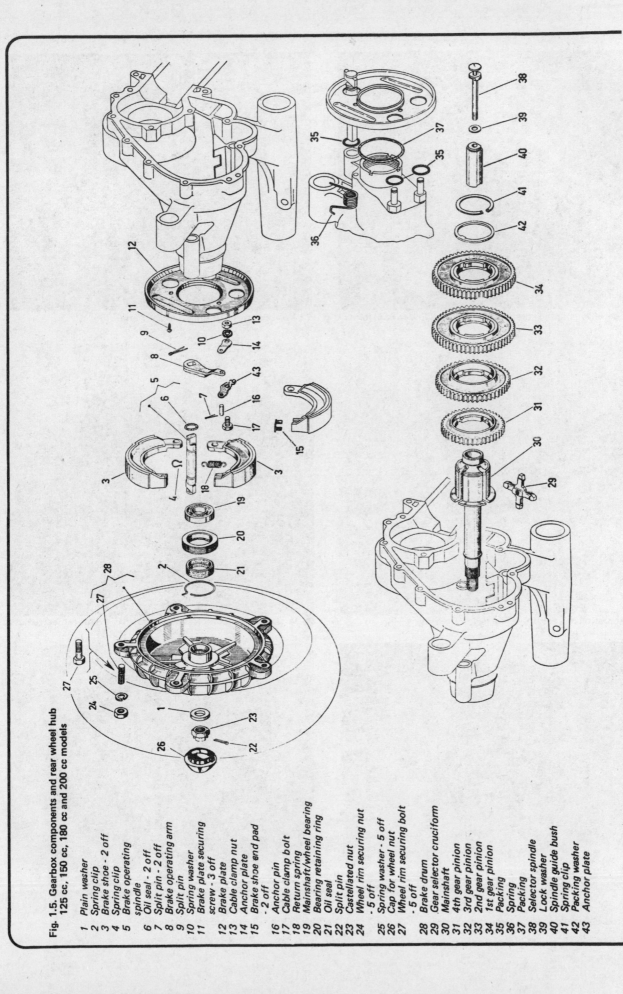

Fig. 1.5. Gearbox components and rear wheel hub
125 cc, 150 cc, 180 cc and 200 cc models

1   Plain washer
2   Spring clip
3   Brake shoe - 2 off
4   Spring clip
5   Brake operating
    spindle
6   Oil seal - 2 off
7   Split pin - 2 off
8   Brake operating arm
9   Split pin
10  Spring washer
11  Brake plate securing
    screw - 3 off
12  Brake plate
13  Cable clamp nut
14  Anchor plate
15  Brake shoe end pad
    - 2 off
16  Anchor pin
17  Cable clamp bolt
18  Return spring
19  Mainshaft/wheel bearing
20  Bearing retaining ring
21  Oil seal
22  Split pin
23  Castellated nut
24  Wheel rim securing nut
    - 5 off
25  Spring washer - 5 off
26  Cap for wheel nut
27  Wheel rim securing bolt
    - 5 off
28  Brake drum
29  Gear selector cruciform
30  Mainshaft
31  4th gear pinion
32  3rd gear pinion
33  2nd gear pinion
34  1st gear pinion
35  Packing
36  Spring
37  Packing
38  Selector spindle
39  Lock washer
40  Spindle guide bush
41  Spring clip
42  Packing washer
43  Anchor plate

20.5 Use valve grinding compound to lap-in cylinder head joint

22.1 Rubber stop cushions quadrant return

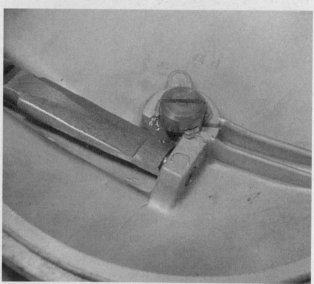

24.1 Actuating mechanism rarely needs attention

25.1a Use vice to compress clutch so that circlip can be removed

25.1b The dismantled clutch assembly

25.2 Order of clutch plate assembly

of carbon behind the rings.

7   It cannot be over-emphasised that the condition of the piston and rings in a two-stroke engine is of prime importance, especially since they control the opening and closing of the ports in the cylinder barrel by providing an effective seal. A two-stroke engine has only three working parts, one of which is the piston. It follows that the efficiency of the engine is very dependent on the condition of this component and the parts with which it is closely asociated.

8   If correctly gapped, the piston rings should have an end clearance within the range 0.006 - 0.008 inch when inserted in the cylinder bore.

### 19   Small end - examination and replacement

1   The small end bearing of a two-stroke engine is more prone to wear, producing the characteristic rattle which is heard in many engines which have covered a considerable mileage. The gudgeon pin should be a good sliding fit in the bearing, without evidence of any play. If play is apparent, the bearing must be renewed.

2   Vespa scooters employ a needle roller bearing assembly which is easy to remove and to renew. It is a light sliding fit in the eye of the connecting rod.

### 20   Cylinder head - examination and renovation

1   It is unlikely that the cylinder head will require any special attention apart from removing the carbon deposit from the combustion chamber. Finish off with metal polish; a polished surface will reduce the tendency for carbon to adhere and will also help improve the gas flow.

2   Ensure that the cooling fins are not obstructed and that they receive the full air flow. A wire brush provides the best means of cleaning.

3   Check the condition of the thread where the spark plug is inserted. The thread in an aluminium alloy cylinder head is damaged very easily if the spark plug is overtightened. If necessary, the thread can be reclaimed by fitting what is known as a Helicoil insert. Most agents have facilities for this type of repair, which is not expensive.

4   If the cylinder head joint has shown signs of oil seepage when the machine was in use, check whether the cylinder head is distorted by laying it on a sheet of plate glass. Severe distortion will necessitate a replacement head but if the distortion is only slight it is permissible to wrap some emery cloth (fine grade) around the sheet of glass and rub down the joint using a rotary motion, until it is once again flat. The usual cause of distortion is uneven tightening of the cylinder head nuts.

5   There is no gasket at the cylinder head to cylinder barrel joint. If the stepped jointing surface has shown a tendency to weep oil, it is permissible to grind the joints in, using coarse grinding paste on the outer joint and fine grinding paste on the inner joint. Grind with a semi-rotary motion, turning backward and forward. Raise the cylinder head occasionally and locate it in a different position before continuing the grinding operation. This will help distribute the grinding paste more evenly. Do not grind more than is necessary to achieve a good mating surface at both joints and make sure all traces of the grinding paste are removed, for this compound is highly abrasive.

### 21   Crankcases - examination and renovation

1   Inspect the crankcases for cracks or any other signs of damage. If a crack is found, specialist treatment will be required to effect a satisfactory repair.

2   Clean off the jointing faces, using a rag soaked in methylated spirit to remove old gasket cement. Do not use a scraper because the jointing surfaces are damaged very easily. A leak-tight crankcase is an essential requirement of any two-stroke engine. Check the bearing housings, to make sure they are not

25.5 Clutch springs seat within depressions in pressure plate

25.6 Reassemble, pressure plate uppermost

28.1a Fit return spring to quadrant first

damaged. The entry to the housings should be free from burrs or lips.

3 Examine the inside wall of the right hand crankcase very carefully, for unless the surface is flat and free from imperfections or blemishes the rotary valve will not function correctly. There is only an 0.002 inch clearance between the outer cheek of the right hand flywheel and the crankcase wall which must be filled with a film of oil if a gas tight seal is to be formed. If the inner surface of the crankcase is damaged or scored, both crankcase halves must be renewed as a matched pair.

## 22 Gearbox components - examination and renovation

1 Examine carefully the gearbox components for signs of wear or damage such as chipped or broken teeth on the gear pinions and kickstarter quadrant, rounded dogs on the ends of the gear pinions or the cruciform selector, weakened or damaged springs and worn splines. If there is any doubt about the condition of a part, it is preferable to play safe and renew the part at this stage. Remember that if a suspect part should fail later, it will be necessary to completely strip the engine/gear unit yet again.

2 It is advisable to renew the kickstarter return spring irrespective of whether it seems to be in good condition. This spring is in constant use, yet if it has to be replaced at a later date, a certain amount of dismantling is necessary in order to gain access. It is cheap and easy to refit at this stage.

3 Do not forget to examine the kickstarter ratchet assembly. Examination will show whether the ratchet teeth have worn, causing the kickstarter to slip or whether the outer teeth are damaged, causing the kickstarter quadrant to jam. Note that the leading tooth of the quadrant is relieved, to help offset the tendency to jam during the initial engagement.

4 The only gearbox bearings left in position are the mainshaft caged roller bearing in the left hand crankcase, and the ball journal bearing in the centre of the layshaft gear cluster which is retained by a circlip. Drive the bearings out of position so that they can be washed in a petrol/paraffin mix and examined, along with the left hand mainshaft ball journal bearing which is still within the left hand crankcase. The bearings should be renewed if they show any signs of play or if any roughness is detected when they are rotated.

5 In the case of the uncaged needle roller bearing used to support the left hand end of the layshaft within the end of the gear cluster, it will be necessary to wash and temporarily reassemble this bearing to check for general wear. In most cases, renewal of all 23 rollers will take up any small amount of play which has developed, especially if the rollers have worn through scuffing on each other. The extreme end of the shaft is supported in a blind bush which is a shrink fit in the left hand crankcase casting. If it is necessary to remove this bush, use a tap to thread into the bush and then heat the crankcase so that the bush can be pulled out by the tap. Fit the new bush whilst the crankcase is still hot, and if necessary, ream out to the correct diameter when the crankcase is cool.

## 23 Multiple gear assembly - examination and renovation

1 The multiple gear assembly (layshaft) has a shock absorber incorporated within the centre of the large diameter helical gear pinion which is driven by the smaller pinion on the end of the crankshaft. It contains a spring-loaded assembly which transmits the drive to the gear train and absorbs any transmission surges. In the smaller capacity models, the shock absorber may be contained within the clutch drive pinion.

2 After a lengthy period of service the springs weaken and the drive becomes less positive. In order to renew the springs, it is necessary to grind off the heads of the rivets that secure the side plates together and punch the rivets out of position so that the side plates can be separated. All of the springs should be renewed. This is an opportune time to renew the gear pinion too, if any wear or damage is evident.

3 Reassemble using the new springs and rivet the side plates

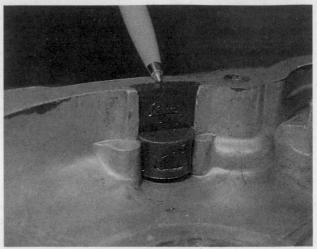
28.1b Renew rubber stop if not in perfect condition

28.2 Quadrant correctly located

29.1a Warm casting before replacing bearing

29.1b Tighten bearing retainer fully

29.3 Grease will hold uncaged rollers in position

29.5a Drive mainshaft through bearing centre

29.5b Insert cruciform through longest slot in mainshaft

29.5c Flats in centre of cruciform must face upwards

29.6 Grease guide bush prior to replacement

29.7a  Assemble pinions, commencing with smallest diameter

29.7b  Fit tongued washer, followed by . . . .

29.7c  . . . . circlip that retains assembly in position

29.8a  Warm crankcase before inserting crankshaft assembly

29.8b  Tap lightly to locate bearing in housing

29.9a  Use thin layer of gasket cement between crankcases

back into position so that the shock absorbing action of the pinion is restored.

## 24 Clutch actuating mechanism - examination

1  The clutch actuating mechanism is contained within the domed, circular cover on the left hand side of the engine unit. It is unlikely that these parts will require attention unless the return spring has weakened or the operating arm itself is stiff to rotate.
2  If it is necessary to release the operating arm for examination or greasing, release the return spring from the stop on the end of the shaft and then withdraw the shaft itself.

## 25 Clutch assembly - examination and renovation

1  To dismantle the clutch assembly, place it in a vice as shown, using a distance piece over the centre. Compress the centre with the vice jaws until the circlip in the back of the housing can be displaced. Slacken the vice gently and the clutch will separate into its component parts.
2  Examine the condition of the linings of the inserted clutch plates. If they are damaged, loose or have worn thin, new linings will be required.
3  Examine the tongues of the plain clutch plates, where they engage with the clutch drum. After an extended period of service, burrs will form on the edges of the tongues which will correspond with grooves worn in the clutch drum slots. These burrs must be removed, by dressing with a smooth file.
4  The grooves worn in the clutch drum slots can be dressed in a similar manner, making sure that the edges of the slots are square once again. If this simple operation is overlooked, clutch troubles will persist because the plates tend to lodge in the grooves when the clutch is withdrawn and promote clutch drag.
5  Check the condition of the clutch springs. They should be replaced if they have compressed.    must be replaced if they
6  Reassemble in the reverse order, using a vice to compress the springs whilst the circlip is reinserted. Note that the end plate against which the circlip abuts is the one which has inserts on one side only.

## 26 Mainshaft/final drive oil seal - examination and renovation

1  There is an oil seal within the centre of the locking ring which retains the ball journal bearing in the left hand crankcase. The oil seal should be displaced and rejected, irrespective of its condition.
2  Fit a new oil seal in the centre of the locking ring, taking care that it is not damaged during fitting. A good seal is essential at this point, otherwise oil from the gearbox will leak along the driveshaft onto the rear wheel.

## 27 Reassembly - general

1  Before the engine, clutch and gearbox components are reassembled, they must be cleaned thoroughly so that all traces of old oil, sludge, dirt and gaskets are removed. Wipe each part clean with a dry, lint-free rag to make sure that there is nothing to block the internal oilways of the engine.
2  Lay out all the spanners and other tools likely to be required so that they are close at hand during the reassembly sequence. Make sure the new gaskets and oil seals are available - there is nothing more infuriating than having to stop in the middle of a reassembly sequence because a gasket or some other vital component has been overlooked.
3  Make sure the reassembly area is clean and unobstructed and that an oil can with clean engine oil is available so that the parts can be lubricated before they are reassembled. Refer back to the torque wrench settings and clearance data where necessary. Never guess or take a chance when this data is available.
4  Do not rush the reassembly operation or follow the instructions out of sequence. Above all, do not use excess force

when parts will not fit together correctly. There is invariably good reason why they will not fit, often because the wrong method of assembly has been used.

## 28 Engine and gearbox reassembly - replacing the kickstarter mechanism

1  Place a new rubber seal in the boss of the right hand crankcase casting, through which the kickstarter shaft protrudes. There is a groove machined to retain this seal. Insert the kickstarter quadrant, after first checking to ensure that the rubber buffer which acts as the quadrant stop is in good order. Lightly grease the shaft to prevent damage to the seal as it is inserted.
2  Wedge the kickstarter quadrant to prevent it moving, then fit and tension the return spring. The outer tang of the spring should locate with the hole drilled in the crankcase.

## 29 Engine and gearbox reassembly - replacing the gearbox components

1  Replace the bearing in the left-hand crankcase, through which the gearbox mainshaft passes, to form the attachment point for the rear wheel. Refit the bearing locking ring (noting that it has a left-hand thread) after fitting a new oil seal within its centre.
2  Tighten the locking ring fully, then replace the wire clip which locks it in position.
3  Grease the roller track of the gearbox layshaft at the right hand end and assemble the 23 uncaged roller bearings in position. Grease the inside right hand end of the layshaft gear cluster and insert the layshaft in the gear cluster, making sure none of the rollers are displaced. Tap the layshaft through the ball journal bearing at the left hand end of the gear cluster.
4  Insert the layshaft complete with gear cluster in the bush of the left hand crankcase and replace the tab washer, plain washer and nut on the end of the shaft which protrudes through the crankcase. Do not tighten fully until the crankcases have been reassembled, at a later stage.
5  Replace the mainshaft through the ball journal bearing in the left hand crankcase. Insert the cruciform through the longest slot in the mainshaft, with the flats in the centre uppermost. Note that the cruciform legs with the rounded edges should be entered in the slot, so that the cruciform can be tilted until it is located correctly.
6  Insert the gear selector guide bush which should be lightly coated with grease. Make sure it locates correctly with the flats on the cruciform, then refit and tighten the gear selector rod, tab washer and locknut. This has a **left hand** thread and must be tightened fully before the tab washer is bent over.
7  Add the individual gear pinions, commencing with the smallest diameter first. The radiused bosses should face outward during assembly. When the final pinion is fitted, add the tongued packing washer, then refit the spring circlip which retains the complete assembly in position.
8  Grease the crankshaft oil seal in the left hand crankcase and warm the crankcase before tapping the crankshaft assembly into position, complete with main bearings. Avoid using force to drive the crankshaft assembly into position and do not overheat the crankcase so that the oil seal deteriorates.
9  Fit the small kickstarter pinion on the gearbox layshaft. Warm the right hand crankcase and coat the jointing faces of both crankcases with a thin layer of gasket cement before fitting them together. Note that it will be necessary to temporarily fit the kickstarter lever so that the kickstarter quadrant can be depressed during the fitting operation. If this precaution is not observed, it will not be possible to join the two cases together due to the quadrant fouling a crankcase projection. Tap the cases gently until they are in firm contact with each other round the periphery, then refit and tighten the twelve retaining bolts, nuts and washers.

## 30 Engine and gearbox reassembly - refitting the clutch

1  Position the Woodruff key in the left hand end of the

29.9b Temporarily fit kickstarter so that quadrant is depressed whilst cases are pressed together

30.1a Fit woodruff key in crankshaft ...

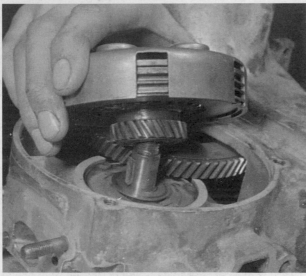

30.1b ... and lower assembled clutch unit into position

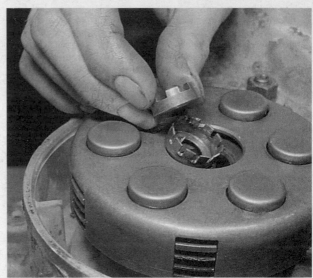

30.1c Fit tab washer and tighten centre nut with peg spanner

30.2 Replace thrust plate in centre

30.3 Make sure domed cover has new 'O' ring seal

crankshaft and lower the assembled clutch into position so that the driving pinion on the back engages with the large diameter pinion of the gearbox layshaft. Refit the tab washer, then the sleeve nut. Tighten the sleeve nut fully, and bend the tab washer to lock it in position.

2  Replace the circular thrust plate in the centre of the clutch, making sure it is retained in position by the wire clip.

3  Grease the clutch actuating pad and attach it to the operating arm within the circular clutch cover. Fit the cover and tighten the three retaining bolts which should each have a spring washer and a plain washer. Gasket cement should not be necessary unless the jointing faces are damaged. The O ring normally makes an effective seal.

4  Do not omit to tighten the adjacent gearbox layshaft nut which has been left slack until the stage of reassembly.

## 31  Engine and gearbox reassembly - refitting the stator coil assembly and flywheel rotor

1  Feed the cables from the stator coil assembly through the passageway in the flywheel housing. The stator assembly is retained by three screws and washers, located around the periphery of the mounting plate. The scribe marks made during the dismantling operation will ensure the baseplate is positioned correctly so that the ignition timing is not disturbed. Then tighten the three screws fully when the scribe lines mate up exactly.

2  Replace the flywheel rotor after first removing the centre circlip, washers and centre nut. Do not omit to insert the Woodruff key in the crankshaft first which ensures that the flywheel is located correctly. Then replace the washers, nut and circlip, making sure the latter seats in its retaining groove.

3  Jar the flywheel nut tight and check that the crankshaft revolves quite freely and has no end float.

## 32 Engine and gearbox reassembly - refitting the piston, cylinder barrel and cylinder head

1  Pad the mouth of the crankcase with rag to prevent any displaced parts from dropping in. If this precaution is overlooked, it may be necessary to separate the crankcases again to retrieve any lost parts.

2  Warm the piston, replace the caged needle roller bearing in the small end of the connecting rod, and flood the roller bearing with oil. Lower the piston into position and press the gudgeon pin into its correct location, preferably with one of the circlips in the piston boss. The piston will have been marked inside the skirt during the dismantling operation, to make sure that it is replaced in its original position.

3  Refit the second circlip and check that **both** are located correctly in their respective grooves. This is most important because a displaced circlip will score the cylinder walls and may cause an engine seizure.

4  Check again that the deflector of the piston is pointing in the correct direction. If the piston is replaced in reverse, a surprising loss of power will be experienced, coupled with a high rate of fuel consumption. Fit a new cylinder base gasket.

5  A piston ring clamp will make the refitting of the cylinder barrel easier, but if one is not available, the piston rings can be fed into the bore by hand, one at a time. The end of the cylinder bore has a lead-in taper to help. Make sure the ends of the piston rings engage with their respective pegs, otherwise difficulty in fitting will be experienced, together with the risk of ring breakage. Oil both the cylinder bore and the piston ring surfaces prior to fitting.

6  When both rings have engaged with the bore correctly, remove the rag from the crankcase mouth and unclip the piston ring clamp. Lower the cylinder barrel into position until it seats on the crankcase mouth.

7  Refit the cylinder head with the spark plug pointing to the rear and tighten down the four retaining nuts and washers evenly, in a diagonal sequence to avoid distortion. **There is no cylinder head gasket.**

31.1 Feed stator plate wires through housing first

31.2 Replace rotor after removing centre nut and circlip

32.2 Gudgeon pin should press into position

32.3 Always fit new circlips and make sure they locate correctly

32.4 Double-check piston deflector faces in correct direction

32.5 Piston ring clamp makes refitting easier

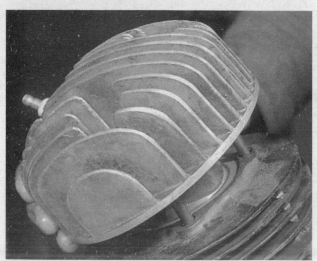

32.6 Spark plug points towards rear of machine

**33   Engine and gearbox reassembly - replacing the cooling fan, cover and cylinder cowl**

1   Before bolting the cooling fan to the flywheel rotor, a check should be made to ensure that the ignition coil is in position. The bolts which retain the coil mounting clamp pass through the back of the ducting casting and cannot be inserted after the cooling fan is in position. Make sure that the bolts are tight and have spring washers fitted. The coil must point toward the cylinder.

2   The cooling fan impeller bolts to the centre of the flywheel generator rotor with four bolts, each fitted with a shakeproof washer. Make sure that these bolts are tight.

3   Refit the fan cover, which is retained to the casting by four screws and washers, around the periphery.

4   Fit and tighten the long sleeve nut which is located on the rear left hand cylinder head stud **after** the cylinder head nut and washer have been replaced and tightened. Fit the cylinder cowl and hold it in position with the short bolt which passes through the cowl and locates with the sleeve nut.

**34   Engine and gearbox reassembly - replacing the rear brake plate and brake shoes**

1   Before replacing the backplate of the brake assembly, first check that the wire retaining clip has been replaced to stake the wheel bearing locking ring securely. It is also advisable to check that the brake operating cam moves freely within its housing at the same time, since it is easy to remove, grease and replace at this stage. If the action is stiff, withdraw the split pin from the operating arm end and draw out the spindle and cam so that it can be cleaned and greased. Do not omit to replace the split pin after reassembly.

2   Lower the backplate of the brake assembly into position and replace the three self-tapping screws which retain it in position. Refit the brake shoes, fitting the left hand one first. When both are in position refit the spring clip on the common pivot.

**35   Engine and gearbox reassembly - refitting the air cleaner case**

1   It is easiest to refit the air cleaner case whilst the engine unit is out of the frame. A new gasket should be fitted to the joint between the base of the air cleaner and the inlet to the crankcase rotary valve.

2   Lower the case into position and retain it with the single countersunk screw close to the carburettor joint. Take care not to drop the screw into the intake passage, or the engine may have to be dismantled again to retrieve it. The engine unit is now ready for refitting to the chassis.

33.1 Coil must be fitted before cooling fan for access to clamp bolts

33.2 Make sure bolts are tight and have shakeproof washers

33.3 Cover is retained by four screws and washers

33.4 Sleeve nut fits over rear left-hand cylinder head stud

34.1a Check wire retaining clip is correctly located

34.1b Grease brake operating cam now if attention is needed

34.2a Back plate is retained by three self-tapping screws

34.2b Fit left-hand brake shoe first

34.2c Don't forget wire clip around brake shoe pivot

35.2 Take care screw does not drop into inlet passage of engine

36.2a Fit front engine mounting bolt first

36.2b Support rear of machine whilst inserting bolts

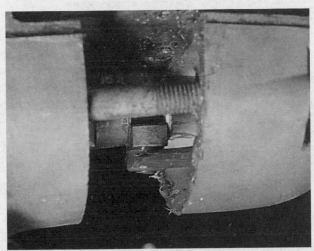

36.3 Make sure selector engages with end of selector rod

36.4 Check all gears select correctly before replacing metal shroud

### 36 Replacing the engine and gearbox unit in the chassis

1 Raise the chassis so that the engine unit can be inserted without clearance problems. It is best to place a billet of wood under the centre stand and to prop up the extreme end of the rear mudguard.

2 Refit the front engine bolt first. This passes right through the chassis, immediately to the rear of the footboards. It can then be used as a pivot for lifting the engine unit whilst the rear engine bolt is refitted. This passes through the bottom of the rear suspension unit. Make sure both bolts have washers beneath the nuts before they are tightened fully.

3 Refit the gearchange mechanism which has been left suspended by its cables. The operating arm must engage with the slot in the selector rod. Select top gear to make assembly easy, then use the handlebar grip to change down as the selector mechanism is fed onto the two retaining studs. The gasket at the joint between the gearbox and the selector mechanism must be in good condition to obviate the risk of oil leaks.

4 When the mechanism has been pushed home, replace the spring washers, plain washers, then fit, and tighten, the two securing nuts. Check that all four gears can be selected correctly, then replace the metal shroud which protects the exposed mechanism. It is retained by a single screw and spring washer.

5 Refit the rear wheel whilst there is still clearance to insert it under the chassis. It is retained by a single castellated nut and washer, which must be tightened fully before the split pin is inserted and bent over. Remove the prop from the rear of the machine so that it stands firmly on the centre stand.

6 Reconnect the clutch and rear brake operating cables on the underside of the machine. Adjustment may be necessary before either control will operate correctly. Both have a cable adjuster; coarse brake adjustment is first obtained by means of the cable clamp on the end of the rear brake operating arm. **This must be tightened fully when adjustment is correct.** Failure to do so may render the brake inoperative under heavy braking.

### 37 Replacing the carburettor, air cleaner and the electrical junction box

1 Replace the carburettor as a complete unit, using a new base gasket It is retained by two long screws which pass through the carburettor casting. No gasket cement is needed at the inlet joint.

36.5a Rear wheel will slide onto splined shaft

36.5b Split must be fitted after castellated nut is tightened fully

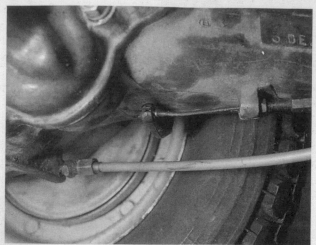

36.6 Cable adjusters are on underside of engine unit

37.2 Reconnect carburettor control cables

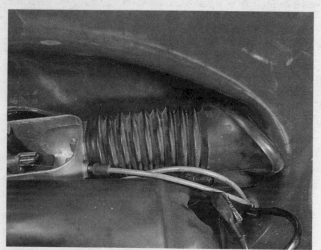

37.4 Air hose acts as air intake for air cleaner box

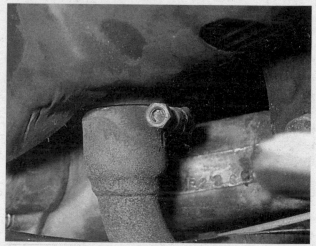

38.1 Tighten exhaust pipe clamp first

2   Reconnect the carburettor control cables with their respective controls. The throttle cable connects with the wire clip which extends from the mixing chamber, closest to the edge of the air cleaner box. The choke lever connects with a slot in the choke operating mechanism.

3   Reconnect the petrol feed pipe which passes through a rubber grommet in the base of the air cleaner box. It connects by means of a banjo union to the carburettor float chamber.

4   Connect the convoluted rubber hose to the stub of the air cleaner box which links the air intake with the air space inside the chassis. Make sure the hose is not kinked or split.

5   Refit the air cleaner element which is retained by two shouldered screws. The carburettor slow running screw will project through the element. Finally, replace the cover which encloses the whole air cleaner assembly, making sure the sealing gasket is still in good order. The cover is retained by two screws.

6   Reconnect the electrical leads from the coil and the flywheel generator with those which extend from the chassis wiring. There is little risk of misconnection if the colour coding is followed throughout, all like colours being interconnected. Fit the junction box cover and retaining screw which holds the closed box to the crankcase casting.

7   Make sure all the coil connections are correct. Push the plug lead into the centre of the coil and make sure that all the water-proofing caps are in good condition and located correctly.

8   Refit the kickstarter pedal. Marks made previously will ensure that it is replaced on the splines in the correct position.

38.3 Make sure gearbox is refilled before replacing filler plug

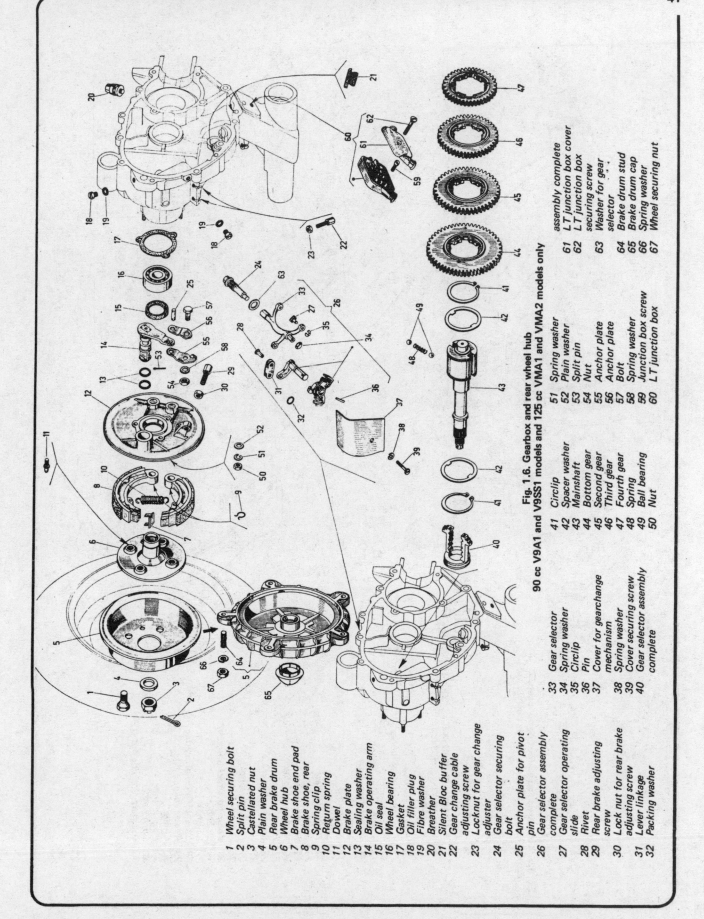

Fig. 1.6. Gearbox and rear wheel hub

90 cc V9A1 and V9SS1 models and 125 cc VMA1 and VMA2 models only

1   Wheel securing bolt
2   Split pin
3   Castellated nut
4   Plain washer
5   Rear brake drum
6   Wheel hub
7   Brake shoe end pad
8   Brake shoe, rear
9   Spring clip
10  Return spring
11  Dowel
12  Brake plate
13  Sealing washer
14  Brake operating arm
15  Oil seal
16  Wheel bearing
17  Gasket
18  Oil filler plug
19  Fibre washer
20  Breather
21  Silent Bloc buffer
22  Gear change cable
    adjusting screw
23  Locknut for gear change
    adjuster
24  Gear selector securing
    bolt
25  Anchor plate for pivot
    pin
26  Gear selector assembly
    complete
27  Gear selector operating
    slide
28  Rivet
29  Rear brake adjusting
    screw
30  Lock nut for rear brake
    adjusting screw
31  Lever linkage
32  Packing washer

33  Gear selector
    complete
34  Spring washer
35  Circlip
36  Pin
37  Cover for gearchange
    mechanism
38  Spring washer
39  Cover securing screw
40  Gear selector assembly
    complete

41  Circlip
42  Spacer washer
43  Mainshaft
44  Bottom gear
45  Second gear
46  Third gear
47  Fourth gear
48  Spring
49  Ball bearing
50  Nut

51  Spring washer
52  Plain washer
53  Split pin
54  Nut
55  Anchor plate
56  Anchor plate
57  Bolt
58  Spring washer
59  Junction box screw
60  LT junction box

61  LT junction box cover
62  LT junction box
    securing screw
63  Washer for gear
    selector
64  Brake drum stud
65  Brake drum cap
66  Spring washer
67  Wheel securing nut

assembly complete

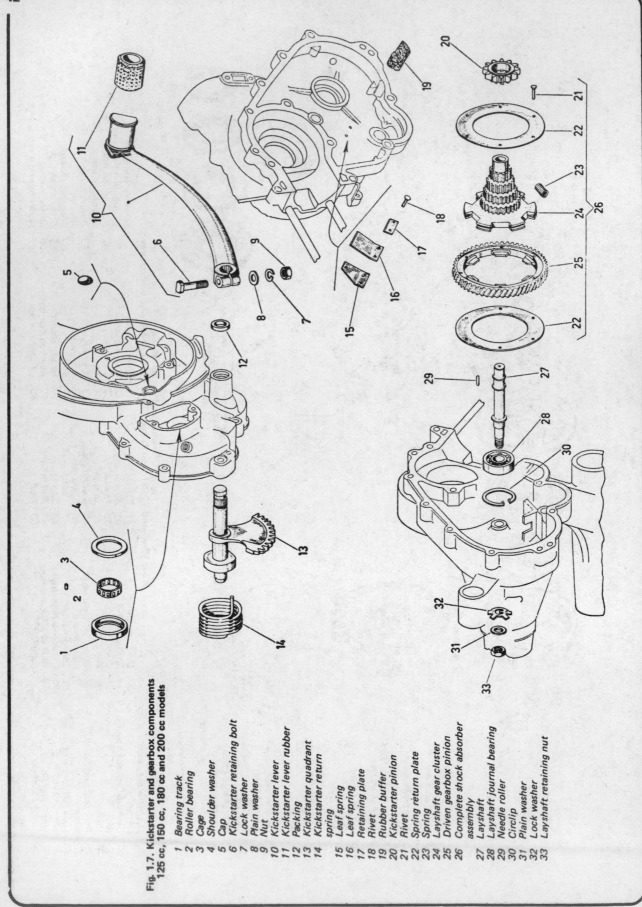

**Fig. 1.7. Kickstarter and gearbox components 125 cc, 150 cc, 180 cc and 200 cc models**

1 Bearing track
2 Roller bearing
3 Cage
4 Shoulder washer
5 Cap
6 Kickstarter retaining bolt
7 Lock washer
8 Plain washer
9 Nut
10 Kickstarter lever
11 Kickstarter lever rubber
12 Packing
13 Kickstarter quadrant
14 Kickstarter return spring
15 Leaf spring
16 Leaf spring
17 Retaining plate
18 Rivet
19 Rubber buffer
20 Kickstarter pinion
21 Rivet
22 Spring return plate
23 Spring
24 Layshaft gear cluster
25 Driven gearbox pinion
26 Complete shock absorber assembly
27 Layshaft
28 Layshaft journal bearing
29 Needle roller
30 Circlip
31 Plain washer
32 Lock washer
33 Layshaft retaining nut

## 38 Replacing the exhaust system and right hand side cover

1 The combined exhaust pipe and expansion box is clamped into position at two points. Push the exhaust pipe on to the exhaust stub of the cylinder first and when it is securely home, tighten the clamp bolt.

2 The silencer box bolts to the underside of the chassis. There is a slotted lug welded to the top of the silencer box which acts as the means of location. Tighten the retaining bolt securely, noting that it should be fitted with both a spring and a plain washer.

3 Before the right hand side cover is fitted, remove the gearbox filler plug, located to the rear of the gear selector mechanism housing. Add sufficient SAE 30 Oil from a pressure oil can until it just commences to run out of the filler plug orifice when the machine is standing on level ground. Let the excess oil drain off, then refit and tighten the plug. Make sure the seating washer of the plug is in good condition to prevent leakage.

4 The right hand side cover pushes into position on the main chassis and is retained by a bonnet-type catch at the forward end.

## 39 Completion of reassembly and final adjustments

1 Reconnect the battery. Check there are no short circuits and that the lights and other electrical equipment work satisfactorily.

2 Check again that the gears select correctly and in their right positions. Select neutral.

3 Check the clutch adjustment is correct and that the carburettor controls operate smoothly.

4 Make sure the tyre pressures are correct, especially if the machine has been left standing for a while. There should be an ample quantity of fuel in the fuel tank, containing the recommended addition of oil.

## 40 Starting and running the rebuilt engine

1 When the initial start-up is made, run the engine slowly for the first few minutes, especially if the cylinder barrel is new or has been rebored. Check that there are no oil leaks. The engine will probably smoke quite heavily to begin with, due to the excess oil within the engine that has accumulated during the assembly process. This should gradually burn away as the engine settles down. It is best to start up in the open air for this reason; never run any engine within a confined space since there is the real danger of carbon monoxide poisoning.

2 Remember that a good seal between the piston and the cylinder barrel is essential for the correct functioning of any two-stroke engine. In consequence, a rebored engine will require more careful running-in than its four-stroke counterpart. There is a far greater risk of engine seizure during the first hundred miles if the engine is permitted to work hard.

3 Do not add extra oil to the petrol/oil mix in the mistaken belief that it will aid running in. More oil means less petrol and the engine will run with a permanently weakened mixture, causing overheating and a far greater risk of engine seizure. Keep to the recommended proportions.

4 Do not tamper with the exhaust system or run the engine without baffles fitted to the silencer. Unwarranted changes in the exhaust system will have a very noticeable effect on engine performance, invariably for the worst.

## 41 Fault diagnosis - engine

| Symptom | Reason/s | Remedy |
| --- | --- | --- |
| Engine will not start | Defective spark plug | Remove plug and lay on cylinder head. Check whether spark occurs when engine is kicked over. |
| | Dirty or closed contact breaker points | Check condition of points and whether gap is correct. |
| | Air leak at crankcase or worn oil seals around crankshaft | Flood carburettor and check whether mixture is reaching the spark plug. |
| | Clutch slip | Check and adjust clutch. |
| Engine runs unevenly | Ignition and/or fuel system fault | Check systems as though engine will not start. |
| | Blowing cylinder head gasket | Leak should be evident from oil leakage where gas escapes. |
| | Incorrect ignition timing | Check timing and reset if necessary. |
| Lack of power | Incorrect ignition timing | See above. |
| | Fault in fuel system | Check system. |
| | Blowing head gasket | See above. |
| | Choked silencer | Clean out or renew. |
| High fuel/oil consumption | Cylinder barrel in need of rebore and o/s piston | Fit new rings and piston after rebore. |
| | Oil leaks or air leaks from damaged gaskets or oil seals | Trace source of leak and renew damaged gaskets or seals. |
| Excessive mechanical noise | Worn cylinder barrel (piston slap) | Rebore and fit o/s piston. |
| | Worn small end bearing (rattle) | Renew bearing and gudgeon pin. |
| | Worn big end bearing (knock) | Fit new big end bearing. |
| | Worn main bearings (rumble) | Fit new journal bearings and seals. |
| Engine overheats and fades | Pre-ignition and/or weak mixture | Check carburettor settings. Check also whether plug grade correct. |
| | Lubrication failure | Is correct measure of oil mixed with petrol? |

## 42 Fault diagnosis - clutch

| Symptom | Reason/s | Remedy |
|---|---|---|
| Engine speed increases but machine does not respond | Clutch slip | Check clutch adjustment for pressure on pushrod. Also free play at handlebar lever. Check condition of clutch plate linings, also free length of clutch springs. Renew if necessary. |
| Difficulty in engaging gears. Gear changes jerky and machine creeps forward, even when clutch is fully withdrawn | Clutch drag | Check clutch adjustment for too much free play. |
| | Clutch plates worn and/or clutch drum | Check for burrs on clutch plate tongues or indentations in clutch drum slots. Dress with file. |
| | Clutch assembly loose on mainshaft | Check tightness of retaining nut. If loose, fit new tab washer and retighten. |
| Operating action stiff | Damaged, trapped or frayed control cable | Check cable and renew if necessary. Make sure cable is lubricated and has no sharp bends. |
| | Bent pushrod | Renew. |

## 43 Fault diagnosis - gearbox

| Symptom | Reason/s | Remedy |
|---|---|---|
| Difficulty in selecting gears | Worn, damaged or bent cruciform | Renew cruciform. |
| Machine jumps out of gear | Gear selector mechanism not adjusted correctly | Adjust cables until gears select correctly at indicated position on handlebar grip. |
| | Chipped or broken teeth on gear pinions | Renew all damaged parts. |
| Kickstarter lever does not return | Broken or badly tensioned return spring | Renew spring or re-tension. |
| Kickstarter jams or slips | Worn quadrant or damaged pinion | Renew defective parts. |

# Chapter 2 Fuel system and lubrication

## Contents

## Specifications

| Model | 312L2 (Sportique) | 232L2 | V9A1 (Vespa 90) | VLA1 (Vespa GL) | VLB1 (Sprint) | V9SS1 (Vespa 90 Super Sport) |
|---|---|---|---|---|---|---|
| Fuel tank capacity | 1.7 Imp gals (7.7 litres) | | 1.15 Imp gals (5.2 litres) | 1.7 Imp gals (7.7 litres) | 1.7 Imp gals (7.7 litres) | 1.23 Imp gals (5.6 litres) |
| **Carburettor** | | | | | | |
| Make | | | Dell'Orto — all models | | | |
| Type | SI 20/17B or CSI 20/15B | | SHB 16/16 | SI 20/17C | SI 20/17D | SHB 16/16 |
| Main jet | 100 | 82 | 63 | 100 | 102 | 82 |
| Slow running (pilot) jet | 42 | 42 | 38 | 42 | 42 | 38 |
| Diffuser | BE/1 or E/1 | E2 | — | E1 | E1 | — |
| Main air bleed | 185 | 150 | — | 140 | 140 | — |
| Throttle valve | No 7 or 0 | No 7 | — | No 0 | No 0 | No 2 |
| Starter jet | 60 | — | 50 | 60 | 60 | 50 |

| Model | VBC1 (Vespa Super) | VMA1 (Vespa 125) | VMA2 (Primavera) | VSD1 (Rally) | V9SS2 (Racer) | VSE1 (Rally 200 Electronic) |
|---|---|---|---|---|---|---|
| Fuel tank capacity | 1.7 Imp gals (7.7 litres) | 1.23 Imp gals (5.6 litres) | 1.23 Imp gals (5.6 litres) | 1.8 Imp gals (8.2 litres) | 1.23 Imp gals (5.6 litres) | 1.8 Imp gals (8.2 litres) |
| **Carburettor** | | | | | | |
| Make | | | Dell'Orto — all models | | | |
| Type | SI 20/15D | SHB 16/16 | SHB 19/19 | SI 20/20D | SHB 16/16 | SI 24/24E |
| Main jet | 88 | 74 | 74 | 109 | 82 | 118 |
| Slow running (pilot) jet | 42 | 42 | 45 | 50 | 38 | 50 |
| Diffuser | E1 | — | — | BE2 | — | BE3 |
| Main air bleed | 160 | — | — | — | — | 160 |
| Throttle valve | — | No 2 | No 2 | — | No 2 | — |
| Starter jet | 60 | 50 | 60 | 60 | 50 | 60 |

**Lubrication**

Petrol/oil mix ... 2% pure mineral oil for all models (¼ pint per 1½ Imp gallons)

## 1 General description

The fuel system comprises a petrol tank under the seat from which a petrol/oil mixture of controlled proportions is fed by gravity to the float chamber of a Dell'Orto carburettor. A petrol tap with a built-in filter bowl is located within the chassis, under the seat. It contains provision for turning on a small amount of reserve fuel when the main content of the tank is exhausted. There is an additional filter of the plastics mesh type, in the top of the float chamber.

For cold starting, the carburettor has a cable-operated choke in the form of a plunger at the side of the mixing chamber. As soon as the engine has been started, the choke can be opened until the engine will accept full air under normal running conditions.

The lubrication system is simple in the extreme. Because the incoming charge is first passed into the crankcase, where it is compressed prior to transfer into the combustion chamber via the transfer ports, oil added to the fuel in a controlled proportion can be used to lubricate the working parts of the engine. This

46

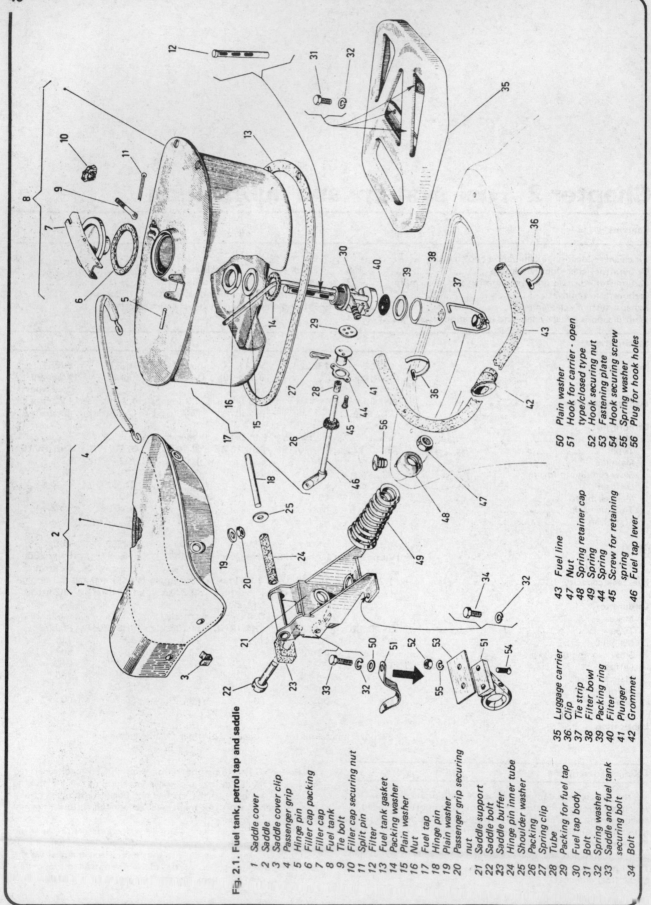

Fig. 2.1. Fuel tank, petrol tap and saddle

1 Saddle cover
2 Saddle
3 Saddle cover clip
4 Passenger grip
5 Hinge pin
6 Filler cap packing
7 Filler cap
8 Fuel tank
9 Tie bolt
10 Filler cap securing nut
11 Split pin
12 Filter
13 Fuel tank gasket
14 Packing washer
15 Plain washer
16 Nut
17 Fuel tap
18 Hinge pin
19 Plain washer
20 Passenger grip securing nut
21 Saddle support
22 Saddle bolt
23 Saddle buffer
24 Hinge pin inner tube
25 Shoulder washer
26 Packing
27 Spring clip
28 Tube
29 Packing for fuel tap
30 Fuel tap body
31 Bolt
32 Spring washer
33 Saddle and fuel tank securing bolt
34 Bolt

35 Luggage carrier
36 Clip
37 Tie strip
38 Filter bowl
39 Packing ring
40 Filter
41 Plunger
42 Grommet
43 Fuel line
47 Nut
48 Spring retainer cap
49 Spring
44 Spring
45 Screw for retaining spring
46 Fuel tap lever

50 Plain washer
51 Hook for carrier - open type/closed type
52 Hook securing nut
53 Fastening plate
54 Hook securing screw
55 Spring washer
56 Plug for hook holes

obviates the need for an oil pump or other means of varying the oil content; as more throttle is used, more petrol and therefore correspondingly more oil is admitted.

It follows that the oil added to the petrol must be of either the self-mixing type, or must be thoroughly dispersed in the petrol to prevent it from settling out or being distributed in uneven quantities. Petroil lubrication does have some disadvantages, but these do not apply to any great extent in engines such as those fitted to the Vespa scooters. The overriding factor is extreme simplicity and freedom from reliance on additional mechanical aids.

### 2 Petrol/oil mix - correct ratio

1 Because the engine relies on the 'petroil' system of lubrication a measured amount of oil must always be present in the petrol. If an oil of the self-mixing type is used, the correct mixing ratio is 2%, or ¼ pint of oil per 1½ gallons of petrol. If a standard (not self-mixing) oil of the recommended viscosity has to be used in an emergency the same mixing ratio still applies, but in this instance the petrol/oil mix must be shaken vigorously until the oil is completely dispersed.

2 The gearbox has its own separate oil content, independent of the engine lubrication system. Two-stroke self-mixing oils must **never** be used.

3 It will be realised that the lubrication of the engine is dependent solely on the intake of fuel mixture from the carburettor. In consequence, it is inadvisable to coast the machine down a long hill with the throttle closed, otherwise there is risk of engine seizure through the temporary lack of lubrication.

4 If the machine is to be laid up for any period, it is advisable to empty the contents of the float chamber by running the engine with the petrol supply turned off. If this precaution is not taken, the petrol in the float chamber will evaporate, leaving behind the oil, which may cause difficult starting on the next occasion.

### 3 Petrol tank - removal and replacement

1 It is unlikely that the petrol tank will need to be removed unless the machine has been laid up and rust has formed on the inside. It may also prove necessary to remove the tank if it leaks, or if there is need to clean out the petrol tap filter bowl.

2 To gain access to the petrol tank, lift the seat and remove the two brackets which hold the nose of the seat and the locking catch at the rear. Each is retained by two bolts. Pull out the grommet from the petrol tap lever so that the lever will pass through the hole in the chassis. The petrol tank can now be lifted out. It is preferable to drain the tank first by detaching the carburettor banjo union within the air cleaner box because the petrol feed pipe must be pulled off the petrol tap assembly before the tank can be lifted clear of the chassis.

3 Note that the petrol tank seats on a gasket within the well of the chassis. This gasket must be in good condition since it cushions the tank from vibration. There should also be a sound gasket within the filler cap to prevent spillage whilst the machine is in motion.

4 When replacing the tank, make sure it seats correctly on the well gasket before the holding-in clamps and bolts are replaced, and that the petrol feed pipe is a good push-on fit to prevent leakage.

### 4 Petrol tap and filter - removal and replacement

1 The petrol tap passes through the base of the petrol tank and is secured inside the tank by a large diameter nut. A washer each side of the orifice in the tank provides a petrol-proof seal.

2 To remove the petrol tap, it is necessary to use a flexible wrench, inserted through the filler cap. Vespa Service Tool T0021064, together with component No 11 is recommended for

7.1 Float chamber filter is located under domed end cap

7.2 Float chamber top is retained by two screws

7.3 Plastic float bears direct on float needle

7.5a Carburettor jets are below small 'L' shaped cover

7.5b Location of jets

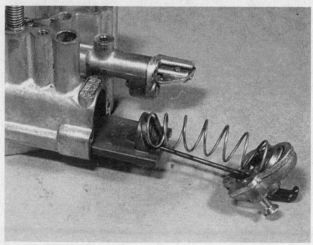

7.6 Throttle slide is of guillotine type

this task.

3   There is no necessity to disturb the petrol tap if only the filter bowl requires attention. To release the bowl, slacken the milled nut at the base so that the wire clip can be displaced to permit removal of the bowl.

4   Remove all sediment from the bowl and clean it thoroughly before replacing. Check that the seating washer for the bowl is in good condition; it should be renewed if there is any doubt.

5   The gauze filter above the filter bowl should be cleaned at the same time. This filter seats **above** the sealing washer.

6   When replacing the petrol tap, use new sealing washers on the outside and inside of the tank. Tighten the retaining nut sufficiently to give a leak-tight joint. The milled nut of the filler bowl assembly should be tightened in similar fashion.

## 5  Petrol feed pipe - examination

1   The petrol feed pipe is made of a synthetic rubber material which may eventually crack or split at the ends, after a long period of service. It is a convenient opportunity to examine the pipe closely when the petrol tank is removed, especially the connection with the petrol tap in the base of the tank.

8.2 Throttle stop and pilot jet adjusting screws

2   Never use ordinary rubber tubing to replace the original because this will rapidly swell and degrade as a result of the action of the petrol. Synthetic rubber tubing which is water-proof must be used or alternatively, plastic tubing in an emergency situation. The latter is not quite so suitable as it will gradually lose its flexibility as a result of the petrol's action.

## 6  Carburettor - removal

1   The carburettor is located within the air cleaner box. To reach it, it is first necessary to remove the lid of the box which is retained by two screws. Detach the air cleaner which is also retained by two screws.

2   Detach the throttle and choke cables from their respective clips on the carburettor. Remove the banjo union from the carburettor float chamber, after first turning off the petrol. Detach the two hexagon headed bolts which pass through the carburettor flange mounting, then lift the complete carburettor away. Note that each of the hexagon headed bolts has a plain washer and a spring washer.

3   There is a gasket between the carburettor flange and the rotary valve inlet in the base of the air filter box which should be renewed every time this joint is broken. Gasket cement should not be necessary.

4   Some models have a different arrangement in which the carburettor is mounted in front of the engine and is connected

to the crankcase by a separate induction pipe. The same rotary induction valve principle is used for the incoming mixture to the engine. In this instance the carburettor is of the clip fitting type and has a mesh-type air cleaner bolted direct to the intake.

## 7 Carburettor - dismantling, examination and reassembly

1 To gain access to the float chamber filter, remove the centre screw from the circular top of the float chamber. There is a sealing gasket under the top cover.
2 The portion of the float chamber which contains the float mounting, float and float needle is retained to the float chamber body by two hexagon headed screws. After removing both screws, lift the top portion of the float chamber away carefully to avoid damaging the gasket or the float assembly.
3 The pin on which the float hinges is a push fit in the mounting. Remove the pin and lift the float out of position. It is made of a plastic material. The float needle can now be lifted out for examination. Wear usually takes the form of a ridge worn around the point of the needle. It may be necessary to use a magnifying glass in view of the small size of this component. If wear is evident, renew the needle.
4 Check also whether the float needle seating is worn. If it is, a new seating should be fitted at the same time.
5 Removal of the small 'L' shaped jet cover in between the float chamber and the mixing chamber will expose the carburettor jets. The cover is retained by a single countersunk screw. The smaller of the two jets is the slow running jet, which should be unscrewed for examination. The larger of the jets is the main air bleed; below it is the diffuser and below that, the main jet. All should be removed for examination.
6 The throttle slide is housed in the mixing chamber. If the two screws which retain the 'D' shaped cap in position are removed, the guillotine-type slide can be withdrawn, together with its return spring. Any signs of wear will immediately be obvious on the flat surfaces of the slide or at the edges. A worn slide should be renewed because it will otherwise have an adverse effect on both petrol consumption and performance.
7 The choke or starting device is housed in a separate compartment on the float chamber side of the mixing chamber. Unscrew the single countersunk screw and lift out the spring-loaded plunger. The starter jet screws into the sidewall of the plunger housing and should be removed at the same time. The carburettor is now dismantled completely with the exception of the pilot jet and throttle stop adjusting screws. If these are removed, note should be made of their positions and the extent of the adjustment, measured from the 'fully home' position, or the carburettor may have to be retuned after reassembly.
8 All jets and passageways of the carburettor should be blown out with compressed air and **never** cleared with wire or any other sharp object. It is otherwise easy to enlarge the jets quite unwittingly and give rise to carburation problems which did not exist previously.
9 During reassembly, never use force to replace any parts. The jets are made of brass and will shear very easily. The carburettor body will not withstand rough usage either; it is cast in a zinc-base alloy and will fracture quite easily if overstressed.
10 Use the dismantling procedure in reverse during reassembly and fit new gaskets wherever possible. It should not be necessary to use gasket cement at any of the joints.

## 8 Carburettor - checking the settings

1 The various jet sizes are predetermined by the manufacturer and should not require modification. Check with the Specifications list to verify whether the values fitted are correct.
2 Slow running is controlled by a combination of the throttle stop and pilot jet settings, irrespective of the type of carburettor

fitted. Commence by screwing inward the throttle stop screw so that the engine runs at a fast tickover speed. Adjust the pilot jet screw until the tickover is even, without either misfiring or hunting. Screw the throttle stop outward again until the desired tickover speed is obtained. Check again by turning the pilot jet screw until the tickover is even. Always make these adjustments with the engine at normal working temperature and remember that the characteristics of a two-stroke engine are such that it is very difficult to secure an even tickover at low engine speeds. Some prefer the engine to stop when the throttle is closed completely, but in a two-stroke engine with petroil lubrication there is always risk of oil starvation if the machine is coasting with the throttle closed.
3 The normal setting for the pilot jet screw is approximately one and a half full turns out from the fully closed position. If the engine 'dies' at low throttle openings, suspect a blocked pilot jet.
4 Guard against the possibility of incorrect carburettor adjustments which result in a weak mixture. Two-stroke engines are very susceptible to this type of fault, which will cause rapid overheating and subsequent engine seizure. Some owners believe that the addition of a little extra oil to the petrol will help prolong the life of the engine, whereas in practice quite the opposite occurs. Because there is more oil the petrol content is less and the engine runs with a permanently weakened mixture!

## 9 Air cleaner - location, examination and cleaning

1 The air cleaner is located inside the air cleaner box, underneath the right hand side cover of the machine. Removal of the lid of the box by removing the two retaining screws will give access to the air cleaner element which is held to the inlet of the carburettor by two shouldered screws.
2 Periodically, as dictated by the routine maintenance schedule, the air cleaner element should be removed from the carburettor, washed in a bath of petrol, then blown dry with a blast of compressed air. No further attention should be necessary.

## 10 Exhaust system - cleaning

1 The exhaust system takes the form of a flat, oblong-shaped expansion box fitted with a short tail pipe, to which the exhaust pipe is attached. It is an integral unit; the exhaust pipe cannot be separated from the expansion box.
2 It is necessary to remove the complete exhaust system before cleaning can take place. First slacken the clamp which holds the exhaust pipe to the exhaust stub of the cylinder barrel, then remove the bolt which passes through the slotted lug on the top of the expansion box. The silencing system is now free from the chassis and can be pulled off the exhaust stub.
3 To clean the system, heat the expansion box externally with a blow lamp and use a piece of hooked wire to free the carbon which is dislodged. It is preferable to undertake this task in the open air, in view of the fumes and smoke which will result. Support the expansion box so that the exhaust pipe hangs downward throughout this operation.
4 The exhaust system should be cleaned out at regular intervals because the accumulation of burnt oil and carbon will otherwise build up and begin to cause back pressures. A blocked or partially blocked expansion box is frequently responsible for a marked fall-off in performance.
5 Before replacing the exhaust system, it is advisable to clean off the outer surface and give it a thick coating of cylinder black or other heat resisting paint to protect the metal from corrosion. Make sure the clamping bolts are tight and that the bolt affixing the expansion box to the underside of the chassis has a plain and a spring washer under the nut, to prevent it working loose.

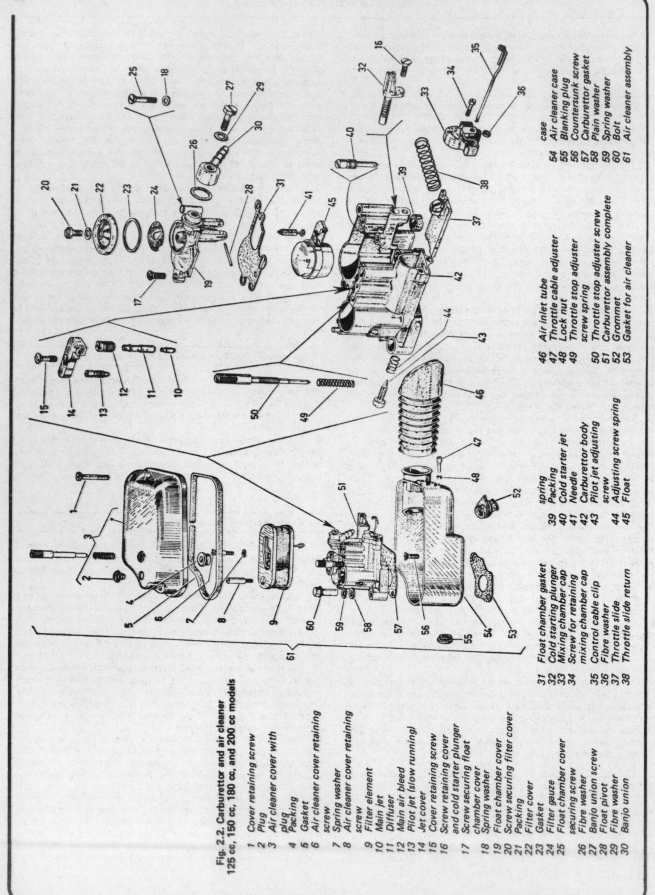

Fig. 2.2. Carburettor and air cleaner
125 cc, 150 cc, 180 cc, and 200 cc models

1  Cover retaining screw
2  Plug
3  Air cleaner cover with plug
4  Packing
5  Gasket
6  Air cleaner cover retaining screw
7  Spring washer
8  Air cleaner cover retaining screw
9  Filter element
10  Main jet
11  Diffuser
12  Main air bleed
13  Pilot jet (slow running)
14  Jet cover
15  Cover retaining screw
16  Screw retaining cover and cold starter plunger
17  Screw securing float chamber cover
18  Spring washer
19  Float chamber cover
20  Screw securing filter cover
21  Packing
22  Filter cover
23  Gasket
24  Filter gauze
25  Float chamber cover securing screw
26  Fibre washer
27  Banjo union screw
28  Float pivot
29  Fibre washer
30  Banjo union
31  Float chamber gasket
32  Cold starting plunger
33  Mixing chamber cap
34  Screw for retaining mixing chamber cap
35  Control cable clip
36  Fibre washer
37  Throttle slide
38  Throttle slide return spring
39  Packing
40  Cold starter jet
41  Needle
42  Carburettor body
43  Pilot jet adjusting screw
44  Adjusting screw spring
45  Float
46  Air inlet tube
47  Throttle cable adjuster
48  Lock nut
49  Throttle stop adjuster screw spring
50  Throttle stop adjuster screw
51  Carburettor assembly complete
52  Grommet
53  Gasket for air cleaner case
54  Air cleaner case
55  Blanking plug
56  Countersunk screw
57  Carburettor gasket
58  Plain washer
59  Spring washer
60  Bolt
61  Air cleaner assembly

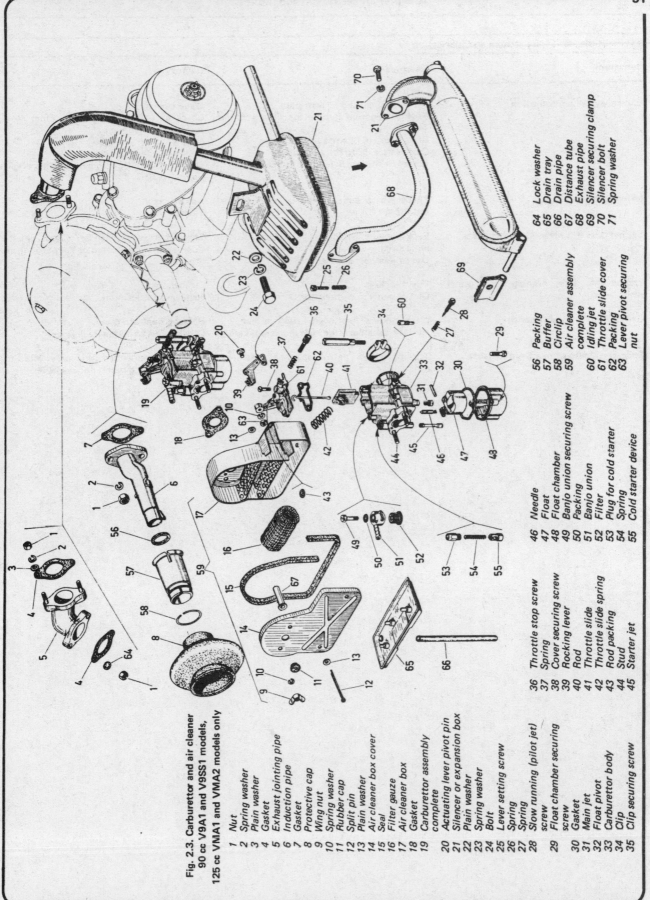

Fig. 2.3. Carburettor and air cleaner
90 cc V9A1 and V9SS1 models,
125 cc VMA1 and VMA2 models only

1  Nut
2  Spring washer
3  Plain washer
4  Gasket
5  Exhaust jointing pipe
6  Induction pipe
7  Gasket
8  Protective cap
9  Wing nut
10  Spring washer
11  Rubber cap
12  Split pin
13  Plain washer
14  Air cleaner box cover
15  Seal
16  Filter gauze
17  Air cleaner box
18  Gasket
19  Carburettor assembly
      complete
20  Actuating lever pivot pin
21  Silencer or expansion box
22  Plain washer
23  Spring washer
24  Bolt
25  Lever setting screw
26  Spring
27  Spring
28  Slow running (pilot jet)
      screw
29  Float chamber securing
      screw
30  Gasket
31  Main jet
32  Float pivot
33  Carburettor body
34  Clip
35  Clip securing screw

36  Actuating lever pivot pin
37  Spring
38  Cover securing screw
39  Rocking lever
40  Rod
41  Throttle slide
42  Throttle slide spring
43  Rod packing
44  Stud
45  Starter jet

46  Needle
47  Float
48  Float chamber
49  Banjo union securing screw
50  Packing
51  Banjo union
52  Filter
53  Plug for cold starter
54  Spring
55  Cold starter device

56  Packing
57  Buffer
58  Circlip
59  Air cleaner assembly
      complete
60  Idling jet
61  Throttle slide cover
62  Packing
63  Lever pivot securing
      nut

64  Lock washer
65  Drain tray
66  Drain pipe
67  Distance tube
68  Exhaust pipe
69  Silencer securing clamp
70  Silencer bolt
71  Spring washer

## 11 Fault diagnosis - fuel system and lubrication

| Symptom | Reason/s | Remedy |
| --- | --- | --- |
| Excessive fuel consumption | Air cleaner choked or restricted | Clean element. |
| | Fuel leaking from carburettor. Float sticking | Check all unions and gaskets. Float needle seat needs cleaning. |
| | Badly worn or distorted carburettor | Renew |
| | Carburettor incorrectly adjusted | Tune and adjust as necessary. |
| | Incorrect silencer fitted to exhaust system | Do not deviate from manufacturer's original silencer design. |
| Idling speed too high | Throttle stop screw in too far. Carburettor top loose | Adjust screw. Tighten top. |
| Engine does not respond to throttle | Back pressure in silencer. Float displaced or punctured | Check baffles in silencer. Check whether float is correctly located or has petrol inside. |
| | Use of incorrect silencer | See above. |
| Engine dies after running for a short while | Fuel blockage | Clean system. |
| | Dirt or water in carburettor | Remove and clean out. |
| General lack of performance | Weak mixture; float needle stuck in seat | Remove float and clean. |
| | Air leak at carburettor joint or in crankcase | Check joints to eliminate leakage. |
| Excessive white smoke from exhaust | Too much oil in petrol, or oil has separated out | Mix in recommended ratio only. Mix thoroughly if mixing pump not available. |

# Chapter 3  Ignition system

## Contents

## Specifications

| Model | | 312L2 (Sportique) | 232L2 | V9A1 (Vespa 90) | VLA1 (Vespa GL) | VLB1 (Sprint) | V9SS1 (Vespa 90 Super Sport) |
|---|---|---|---|---|---|---|---|
| **Spark plugs** | | | | | | | |
| Make | ... | KLG | KLG | KLG | KLG | KLG | KLG |
| Type | ... | F75 | F70 | F80 | F70 | F70 - 75 | F70 |
| Make | ... | Lodge | Lodge | Lodge | Lodge | Lodge | Lodge |
| Type | ... | 2HN | HN | 2HN | HBN | HBN or HN | HBN |
| Make | ... | Champion | Champion | Champion | Champion | Champion | Champion |
| Type | ... | L81 | L86 | L81 | L86 | L86 - L81 | L86 |
| Thread size | | | | 14 mm — all models | | | |
| Reach | ... | ½ in | ½ in | ½ in | ½ in | ½ in | ½ in |
| Gap | ... | | | 0.020 in - 0.025 in — all models | | | |
| **Ignition timing ± 1% BTDC** | ... | 28° | 26° | 19° | 22° | 22° | 19° |

| Model | | VBC1 (Vespa Super) | VMA1 (Vespa 125) | VMA2 (Primavera) | VSD1 (Rally) | V9SS2 (Racer) | VSE1 (Rally 200 Electronic) |
|---|---|---|---|---|---|---|---|
| **Spark plugs** | | | | | | | |
| Make | ... | KLG | KLG | KLG | KLG | KLG | KLG |
| Type | ... | F70 - F75 | F70 - F75 | F70 - F75 | FE80 | F70 - F75 | FE 80 |
| Make | ... | Lodge | Lodge | Lodge | Lodge | Lodge | Lodge |
| Type | ... | HBN - HN | HBN - HN | HBN - HN | 2 HLN | HBN - HN | 2HLN |
| Make | ... | Champion | Champion | Champion | Champion | Champion | Champion |
| Type | ... | L86 - L81 | L86 - L81 | L86 - L81 | NA8 | L86 - L81 | NA8 |
| Thread size | | | | 14 mm — all models | | | |
| Reach | ... | ½ inch | ½ inch | ½ inch | ¾ inch | ½ inch | ¾ inch |
| Gap | ... | | | 0.020 in - 0.025 in — all models | | | |
| **Ignition timing ± 1% BTDC** | ... | 22° | 25° | 24° | 22° | 24° | 24° |

Contact breaker gap ... ... ... ... ... ... ... ...   Within range 0.011 - 0.019 in (0.3 - 0.5 mm) — all models

## 1 General description

A flywheel generator fitted to the end of the crankshaft provides the current for operating the ignition circuit and recharging the battery which forms part of the lighting circuit. Because the output from the generator is AC it is necessary to incorporate a rectifier in the circuit to convert this output to DC so that the battery can be charged.

An external ignition coil is used to generate the spark used to ignite the mixture in the cylinder. The current for this coil is derived from the flywheel generator and in consequence the ignition circuit functions quite independently of the battery.

## 2 Flywheel generator - checking the output

1   The output and general performance of the generator fitted to the Vespa scooter can be checked only with specialised test equipment of the multi-meter type. It is unlikely that the average rider will have access to this type of equipment or instruction in its use. In consequence, if generator performance is suspect, the generator should be checked by a Vespa agent or a qualified auto-electrician. Reduced output is usually caused by de-magnetised magnets in the flywheel rotor. Each magnet should support a weight of 1 - 1½ lbs if in good order. If necessary, the rotor can be remagnetised through the services of a Vespa agent.

2   Failure of the generator does not necessarily mean that a

4.2 Access to contact breaker points for adjustment is through slots in rotor

4.4a Use feeler gauge to check gap when points are fully open

rotor until one or other of these slots coincides with the location of the points in the 1 o'clock position.

3   Rotate the engine until the contact breaker points are in the fully open position. Examine the faces of the contacts. If they are pitted or burnt it will be necessary to remove them for further attention, as described in Section 5 of this Chapter.

4   Adjustment is carried out by slackening the screw which retains the fixed contact point and moving the plate upward or downward (using the eccentric adjustment pin) until the correct gap has been restored. Retighten the screw and check again that the gap is correct. It is important that the contacts are in the fully open position during this check, otherwise a completely false setting will be obtained. If the gap is correct, the feeler gauge should be a good sliding fit.

## 5  Contact breaker points - removal, renovation and replacement

1   If the contact breaker points are burned, pitted or badly worn, they should be removed for dressing. If it is necessary to remove a substantial amount of material before the faces can be restored, the points should be renewed as a set.

2   Before the contact breaker points can be removed, it is necessary to draw off the flywheel rotor using the self-extracting nut in the centre. Refer to Chapter 1, Section 8.3 for the procedure to adopt. Do not fail to bridge the magnets with soft iron immediately the rotor is withdrawn or loss of magnetism will cause a drop in generator output.

3   To release the moving contact, detach the spring clip around the pivot pin and lift the contact off, after slackening the terminal at the end of the return spring. Do not lose the essential insulating washers and take note of their position, to aid subsequent replacement. The plate holding the fixed contact is retained to the baseplate by a single screw.

4   The points should be dressed with an oilstone or fine emery cloth. Keep them absolutely square during the dressing operation, otherwise they will make angular contact when they are replaced and will quickly burn away.

5   Replace the contacts by reversing the dismantling procedure. Take particular care when replacing the insulating washers, to make sure they are fitted in the correct order. If this precaution is not observed the points will be isolated electrically and the ignition system will not function.

6   It is a good opportunity to lightly grease the contact breaker cam and also oil the felt on the moving contact arm when the flywheel rotor is removed.

4.4b Eccentric screw facilitates variation of gap during adjustments

replacement is needed. It is possible to replace the individual ignition and/or lighting coils in the event of their failure.

## 3  Ignition coil - checking

1   The ignition coil is a sealed unit, clamped to the rear of the cooling fan housing. If a weak spark and difficult starting cause its performance to be suspect, it should be tested by a Vespa agent or an auto-electrical mechanic. It is necessary to renew a defective coil because it is not practicable to effect a satisfactory repair.

2   Before suspecting the coil, it is advisable to ensure whether the condenser in the ignition circuit is at fault, because this can give rise to identical symptoms. Refer to Section 6 of this Chapter for the relevant information.

## 4  Contact breaker - adjustment

1   The contact breaker assembly forms part of the flywheel magneto generator and before access to the points can be gained it is necessary to remove the right hand side cover, cylinder cowl, cooling fan cover and the fan impeller. Refer to Chapter 1, Section 8, paragraphs 1 and 2 for the method of dismantling.

2   Access to the contact breaker points is available through two slots in the flywheel rotor and it will be necessary to rotate the

## 6  Electronic ignition

1   The 200 cc Rally Electronic model takes advantage of

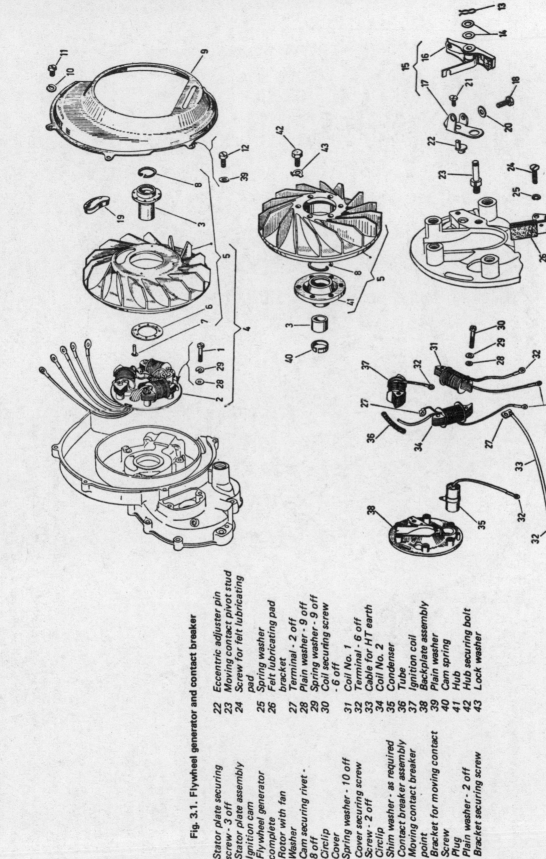

Fig. 3.1. Flywheel generator and contact breaker

1  Stator plate securing screw - 3 off
2  Stator plate assembly
3  Ignition cam
4  Flywheel generator complete
5  Rotor with fan
6  Washer
7  Cam securing rivet - 8 off
8  Circlip
9  Cover
10  Spring washer - 10 off
11  Cover securing screw
12  Screw - 2 off
13  Circlip
14  Shim washer - as required
15  Contact breaker assembly
16  Moving contact breaker point
17  Bracket for moving contact
18  Screw
19  Plug
20  Plain washer - 2 off
21  Bracket securing screw
22  Eccentric adjuster pin
23  Moving contact pivot stud
24  Screw for felt lubricating pad
25  Spring washer
26  Felt lubricating pad bracket
27  Terminal - 2 off
28  Plain washer - 9 off
29  Spring washer - 9 off
30  Coil securing screw - 6 off
31  Coil No. 1
32  Terminal - 6 off
33  Cable for HT earth
34  Coil No. 2
35  Condenser
36  Tube
37  Ignition coil
38  Backplate assembly
39  Plain washer
40  Cam spring
41  Hub
42  Hub securing bolt
43  Lock washer

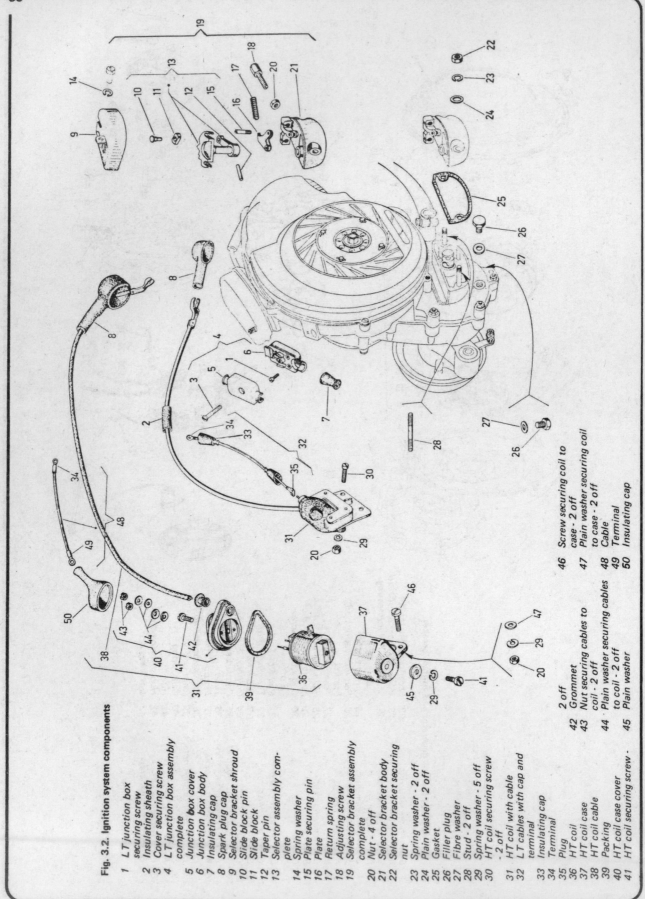

**Fig. 3.2. Ignition system components**

1 LT junction box securing screw
2 Insulating sheath
3 Cover securing screw
4 LT junction box assembly complete
5 Junction box cover
6 Junction box body
7 Insulating cap
8 Spark plug cap
9 Selector bracket shroud
10 Slide block pin
11 Slide block
12 Taper pin
13 Selector assembly complete
14 Spring washer
15 Plate securing pin
16 Plate
17 Return spring
18 Adjusting screw
19 Selector bracket assembly complete
20 Nut - 4 off
21 Selector bracket body
22 Selector bracket securing nut
23 Spring washer - 2 off
24 Plain washer - 2 off
25 Gasket
26 Filler plug
27 Fibre washer
28 Stud - 2 off
29 Spring washer - 5 off
30 HT coil securing screw - 2 off
31 HT coil with cable
32 LT cables with cap and terminal
33 Insulating cap
34 Terminal
35 Plug
36 HT coil
37 HT coil case
38 HT coil cable
39 Packing
40 HT coil case cover
41 HT coil case securing screw - 2 off
42 Grommet
43 Nut securing cables to coil - 2 off
44 Plain washer securing cables to coil - 2 off
45 Plain washer
46 Screw securing coil to case - 2 off
47 Plain washer securing coil to case - 2 off
48 Cable
49 Terminal
50 Insulating cap

modern technology by using electronic circuitry to dispense with the conventional contact breaker assembly and its inherent disadvantages. The current derived from the ignition coil within the flywheel generator is rectified and stored in a condenser until a transistor is triggered to release this charge and pass it to the primary windings of the external ignition coil. High tension current is built up in the secondary windings of the coil and it is this that provides the spark at the spark plug. A cam on the crankshaft determines when the transistor circuit will be triggered as it passes a magnetic pick-up. It is the relationship between the relative positions of the cam and the pick-up which determines the exact point at which the spark will occur.

2 The advantages of this type of system are achieved in the form of a more intense spark, the ability to fire evenly at very high rpm and a continuation of the firing sequence even if the spark plug electrodes become fouled or are badly gapped. Because there are no moving parts in the electronic circuitry no mechanical problems occur and a certain cold start is guaranteed, even if the machine has been laid up for a lengthy period. The electronic components are encapsulated in resin to protect them from both weather and vibration; because the cam does not contact the electrical pick-up, no wear can occur here either. In consequence, no maintenance is necessary.

3 In the unlikely event of the system failing, it will be necessary to seek the assistance of a Vespa repair specialist. It follows that more sophisticated equipment is required when checking an electronic ignition system, especially since the misuse of test equipment may cause irreparable damage to the component parts of the circuit.

## 7 Condenser - removal and replacement

1 A condenser is included in the contact breaker circuitry to prevent arcing across the contact breaker points as they separate. It is connected in parallel with the points and if a fault develops the ignition system will not function correctly.

2 If the engine is difficult to start or if misfiring occurs, it is possible that the condenser has failed. To check, examine the contact breaker points. If they have a blackened or burnt appearance, the condenser can be regarded as unserviceable.

3 It is not possible to check the condenser without the necessary test equipment. In view of the low cost of a replacement it is preferable to change the condenser and observe the effect on engine performance.

4 The condenser is inserted endwise within a boss which forms part of the contact breaker baseplate. It is retained in position by a single screw which passes through a clamp soldered to the condenser body. Disconnect the condenser lead wire, remove the screw and withdraw the condenser. It is necessary to obtain an exact replacement in view of the manner in which the condenser is fitted.

## 8 Ignition timing - checking and resetting

1 If the ignition timing is correct, the contact breaker points will commence to separate when the piston is 28° before top dead centre. Before the timing is checked or re-set, it is essential that the contact breaker gap is verified to be within the range 0.011 - 0.019 inch (0.3 - 0.5 mm) when the points are fully open. Slight variations in ignition timing occur, according to the model. Check with the Specifications Section of this Chapter first.

2 Optimum performance depends on the accuracy with which the timing is set. The use of a degree disc is recommended, which can be bolted to the holes left vacant by removal of the flywheel fan. A Vespa timing disc T0023465 is available for this purpose, together with another service tool T0030259 which will screw into the spark plug hole to form a positive stop for the piston.

3 Attach the timing disc to the flywheel rotor but do not tighten it until it has been zeroed. Attach a pointer to one of the fan cover screws and hold it firmly in this position. A piece of bent wire is especially suitable for this purpose. Remove the spark plug and rotate the engine until the piston is **exactly** at top dead centre. Adjust the pointer and/or the degree disc until the pointer is on the zero mark, then tighten both so that their positions cannot move.

4 Turn the engine **backwards** beyond the 28° mark or whatever other setting is specified, then turn forward again until the pointer registers with the disc at the desired advance reading. If the service tool is inserted **after** the engine has been turned backward, it will act as a stop when the engine is turned forward again. If the timing is correct, the contact breaker points will have commenced to separate.

5 A convenient means of checking whether the points have separated is to insert a piece of cigarette paper between the points when the engine is moved backward and then to turn the engine forward until the grip on the paper is relaxed. Alternatively, a battery and bulb can be used, attaching one lead of the battery to the moving contact breaker arm and the other to a convenient earthing point. If the bulb is incorporated in one of the battery leads, it will be extinguished immediately the points separate.

6 Only a 1° error is permissible. If the reading obtained is outside the limit of error, the flywheel rotor must be withdrawn and the contact breaker baseplate either advanced or retarded until a correct reading is obtained, after the rotor has been replaced. The baseplate is held by three screws passing through slots which permit a limited range of adjustment.

7 Always recheck the timing, especially if the original setting has been changed. A small error can have a surprising effect on performance and fuel consumption.

8 Note the ignition timing is fixed. There is no provision for either advancing or retarding the setting as engine speed increases or decreases.

## 9 Spark plug - checking and resetting the gap

1 A 14 mm short reach spark plug is fitted to the Vespa scooters covered by this manual. Refer to the Specifications Section of this Chapter for the recommended grades.

2 A spark plug with a ½ inch reach is used. Always fit the grade of plug recommended or the exact equivalent in another manufacturer's range. Irrespective of the make of plug fitted, the gap at the electrodes should be 0.020 inch.

3 Check the gap at the plug points every 1000 miles. To reset the gap, bend the outer electrode closer to the central electrode and check that a 0.020 inch feeler gauge can be inserted. Never bend the centre electrode, otherwise the insulator will crack, causing engine damage if particles fall in whilst the engine is running.

4 The condition of the spark plug electrodes and insulator can be used as a reliable guide to engine operating conditions. See accompanying diagrams.

5 Always carry a spare spark plug of the correct grade. The plug in a two-stroke engine leads a hard life and is liable to fail more readily than when fitted to its four-stroke counterpart.

6 Never overtighten a spark plug, otherwise there is risk of stripping the threads from the cylinder head, especially those cast in light alloy. A stripped thread can be repaired by using what is known as a Helicoil insert, a low cost service which is operated by a number of dealers.

7 Use a spanner which is a good fit, otherwise the spanner may slip and break the plug insulator. The plug should be tightened sufficiently to seat firmly on its sealing washer.

8 Make sure the plug insulating cap is a good fit and free from cracks. This cap contains the suppressor which eliminates radio and TV interference.

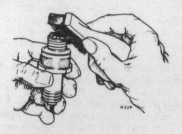
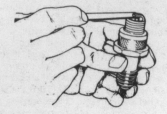

*Cleaning deposits from electrodes and surrounding area using a fine wire brush.*

*Checking plug gap with feeler gauges*

*Altering the plug gap. Note use of correct tool.*

**Fig. 3.3a. Spark plug maintenance**

*White deposits and damaged porcelain insulation indicating overheating*

*Broken porcelain insulation due to bent central electrode*

*Electrodes burnt away due to wrong heat value or chronic pre-ignition (pinking)*

*Excessive black deposits caused by over-rich mixture or wrong heat value*

*Mild white deposits and electrode burnt indicating too weak a fuel mixture*

*Plug in sound condition with light greyish brown deposits*

**Fig. 3.3b. Spark plug electrode conditions**

## 10  Fault diagnosis - ignition system

| Symptom | Reason/s | Remedy |
|---|---|---|
| Engine will not start | No spark at plug | Try replacement plug if gap correct. Check whether contact breaker points are opening and closing, also whether they are clean. Check ignition switch and ignition coil. |
| Engine starts but runs erratically | Intermittent or weak spark | Try replacement plug. Check whether points are arcing. If so, replace condenser. Check accuracy of ignition timing. Low output from flywheel magneto generator or imminent breakdown of ignition coil. |

# Chapter 4 Chassis

## Contents

## 1 General description

A scooter does not use a frame in the generally accepted sense of the word, the chassis taking the form of a one-piece metal pressing which serves as the bodywork as well as providing the points of attachment for the front and rear suspension, the latter of which serves also as the engine unit mounting.

The chassis is of the 'step through' type and has an apron in the front to give the rider good weather protection. The rider sits above the fuel tank, below which is the swinging arm rear suspension with the engine unit mounted on the right hand side so that an extension of the gearbox mainshaft drives the rear wheel direct. Access to the engine unit is usually by means of a detachable 'blister' or side cover; there is a matching side cover on the left hand side which contains the battery and tool kit. This is not readily detachable, a lockable flap providing access to the interior. There are some variations in chassis design; not all have detachable blisters.

Front suspension takes the form of a single strut on the right hand side of the front wheel, attached to the steering head, having a pivot at the bottom on which a trailing link arrangement carries the wheel. Movement is controlled by an hydraulic damper unit acting in conjunction with a heavy duty coil spring. Two types of suspension are in use, both fully described in this Chapter.

Rear suspension takes the form of a swinging arm arrangement, the engine and gearbox unit forming an integral part of the casting. Movement is controlled by a damper unit similar to that adopted for the front wheel; it will be appreciated that with this layout the complete engine and gearbox unit move in unison with swinging arm.

## 2 Front suspension - removal from the chassis

1 It is unlikely that the front suspension will have to be removed from the chassis as a complete unit unless the steering head assembly requires attention or if the machine suffers frontal damage in an accident.
2 To release the front suspension as a complete unit, it is first necessary to detach the handlebars. Commence by removing the centre of the three screws below the headlamp casting. Raise the speedometer from its housing whilst easing the drive cable upward through the steering column. When the speedometer has been raised sufficiently, disconnect the cable by unscrewing the gland nut and detach the cable from the drive. Remove the clip

from the side of the speedometer and the bulb within.
3 Release the front brake cable from the brake by removing the clamp at the operating arm. Draw the cable upward through the steering column after it has been pulled clear of the adjuster. Slacken and remove the clamp bolt below the headlamp casting and lift the handlebar assembly clear. It can be left suspended by the cables, provided they are not permitted to stretch or twist.
4 Raise the scooter so that the front wheel is about one foot from the floor. Remove the front wheel by unscrewing the four wheel nuts and withdraw it from the brake drum.
5 Unscrew the slotted sleeve nut at the top of the steering column. Vespa service tool T0014566 is specified for this purpose, but if it is not available, careful application of a flat nosed punch and hammer can be used to start the nut. Remove the nut and the lockwasher below.
6 The top steering head cone is slotted in similar fashion and can be unscrewed the same way. Take care to support the steering column whilst the cone is slackened and removed, preferably by grasping the front mudguard. When the cone has been removed, the steering column assembly can be withdrawn from the steering head of the chassis, in a downward direction. Note that the ball bearings of the steering head assembly are caged and will not be displaced as the cups and cones separate.

## 3 Dismantling the front suspension assembly

1 If the front suspension assembly requires attention, it is not necessary to remove it from the machine. The procedure detailed in the preceding Section can be by-passed and the dismantling commenced from this point onward, after first removing the front wheel.
2 Detach the speedometer drive cable from the stub axle housing by unscrewing the gland nut. Remove the two countersunk screws from the centre of the brake drum and pull the brake drum off the brake shoes and backplate.
3 Pull the aluminium cover off the suspension link. It may be necessary to use a screwdriver to provide some leverage as the cover is a tight push fit. Unscrew the hexagon headed grease retainer within the link and withdraw the distance piece behind it. Access is now available to the nut retaining the stub axle which can be unscrewed with a ring spanner if the other end of the axle is held firmly to prevent it from rotating during the

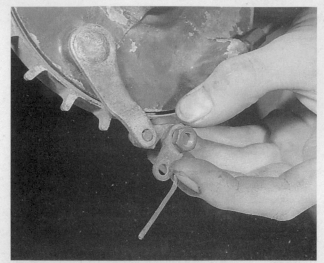

2.3 Clamp acts as cable anchorage

2.4 Remove the four centre nuts to withdraw the front wheel

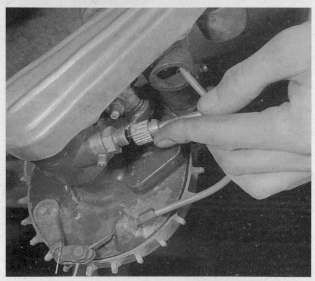

3.2a Detach speedometer drive by unscrewing gland nut

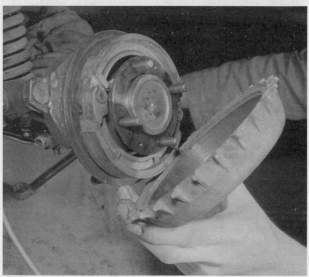

3.2b Two small screws retain brake drum to stub axle

3.3a Aluminium alloy cover will pull off fork link

3.3b Remove grease retainer for access to stub axle nut

3.3c Ring spanner can be used when spacer is removed

3.3d Axle will drive outwards, through bearing centres

3.4 Brake shoes will pull off after pivot clip is removed

3.5 Release lower end of suspension unit by withdrawing bolt

3.6a Pivot pin acts also as brake plate anchorage

3.6b Spring pivot will drive out; acts as pivot for brake shoes too

3.6c Twist pivot to release from end of spring

3.6d Stub axle housing can be lifted away for bearing attention

initial slackening. The axle will drive out of the axle housing after the nut has been removed, with a few taps from a hide mallet.

4   The brake shoes can now be removed from the brake plate by withdrawing the wire clip over their common pivot, then drawing them upward off the operating cam whilst pressure is maintained to keep the cam ends apart. The cam can be withdrawn from the brake plate by pulling it through from the outside.

5   Release the lower end of the front damper unit by removing the retaining nut, spring washer and plain washer. The unit will pull off the bolt through the forward end of the linkage.

6   Slacken and remove the nut and washer on the end of the pivot pin which passes through the end of the steering column. Slacken and remove the nut and washer which secure the end of the suspension spring to the linkage at the same time. The pivot can be driven outward through the lip of the brake plate which acts as an anchorage and stop. Drive the spring pivot outward through the brake plate; it acts also as the pivot for the brake shoes. The stub axle housing is now completely free from the steering column.

7   Some models have a slightly different front suspension layout in which the suspension spring forms part of the front damper assembly. A different dismantling procedure is needed in consequence, which should be carried out as follows:

8   Remove the front wheel by unscrewing the five nuts around the periphery of the wheel rim. Note that each nut has a spring washer below. Disconnect the speedometer drive cable by slackening the retaining plate located between the lower end of the steering column and the damper unit. There is no necessity to remove the bolt completely before the cable can be withdrawn; it is somewhat tricky to replace.

9   Release the front brake cable from the brake operating arm clamp and pull it through the cable adjuster. Prise back the two lugs which retain the cover over the suspension link and remove the cover. A strip rubber seal will be displaced as the outer cover is withdrawn. Unscrew the two nuts within the cover, the larger of which seats on a spring washer. The inner cover can now be withdrawn and the various washers and seals below. Make a drawing of the washers and seals to ensure their correct location on reassembly.

10  The hub assembly can now be detached as a complete unit. If the damper unit requires examination, it can be removed by removing the upper nut where the unit is attached to the steering column bracket.

11  To dismantle the hub, first remove the slotted dome plug and washer which gives access to the speedometer drive pinion. The pinion, made of nylon, will lift out. Remove also the circular rubber plug which blanks off the right hand end of the stub axle. Prise off the cap in the centre of the brake drum and remove the split pin, then the castellated nut which secures the brake drum to the axle. Take off the brake drum, which on some

models is retained by a locating dowel which must be unscrewed first.

12  Move to the right hand side of the hub and unscrew the speedometer drive pinion which is recessed. Hold the wheel flange and turn the nut clockwise, it has a left hand thread. The stub axle can now be driven out using an alloy drift. Extract the oil seal and release the circlip which retains the left hand wheel bearing. The bearing can now be driven out or extracted by the use of Vespa service tool T0029538. Before the right hand bearing can be drifted out, it is necessary to remove the locking ring which secures it in position. The ring has a left hand thread and will need the use of Vespa service tool T0030631 if it is to be removed without damage. The hub is now dismantled with the exception of the brake shoes; these can be removed as described in paragraph 4 of this Section.

13  Reassembly is achieved by reversing these instructions. Check to make sure that all the seals are replaced in their correct order and use a new split pin for securing the castellated nut in the wheel centre.

## 4  Steering head bearings - examination and renovation

1   Before commencing reassembly of the front suspension system, assuming it has been removed as a complete unit, examine the steering head races. The ball bearing tracks of the respective cup and cone bearings should be polished and free from indentations and cracks. If signs of wear or damage are evident, the cups and cones must be renewed.

2   If the cups and cones are renewed, it is advisable to renew both caged ball races at the same time. They are comparatively inexpensive and their renewal will obviate the need for a further complete removal of the front suspension assembly which may otherwise be needed at a later date.

## 5  Front suspension - examination of the component parts

1   If the action of the front damper unit is suspect, it is advisable to detach the unit by removing the nut and bolt from the upper end, where it is bolted to a lug on the extension of the steering column. If the unit rebounds immediately it is compressed and then released, it is no longer functioning correctly. This would cause excessive movement of the front wheel and even front wheel patter.

2   The unit can be dismantled for examination but it is doubtful whether this amount of work is advisable because it is quicker and most probably cheaper in the long run to renew the complete unit. The unit will normally give long service, because it is fully enclosed. A replacement is available on a part-exchange basis.

3.7a Remove five nuts to release front wheel

3.7b Slacken, but do not remove clamp plate bolt

3.8a Release brake cable from operating arm clamp

3.8b Prise back lugs to remove cover

3.8c Make note of location of washers and seals ...

3.8d ... 'O' ring seals are under washers

3.9a Detach hub as a complete unit

3.9b Mirror shows how top of damper unit is attached

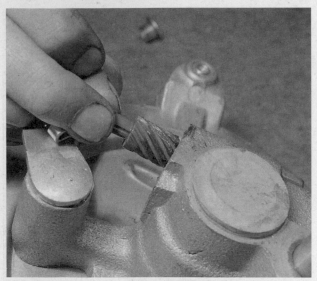

3.10a Speedometer drive pinion will pull out of housing

3.10b Remove castellated nut to release brake drum

3.10c Brake shoes assembly will be exposed

3.11a Rubber plug blanks off end of stub axle

3.11b Speedometer drive pinion nut has left-hand thread

3.11c Use alloy drift when driving axle out of position

3.11d Vespa service tool must be used to unscrew bearing locking ring

3   Check the rubber mounting bushes at the top and bottom of the unit. If they have degraded or show signs of cracking, both should be replaced.

4   The coil spring is unlikely to give trouble, other than breakage in an extreme case. After a lengthy period of service, the spring may compress and take a permanent set. Compare the free length with that of a new spring and if there is any marked difference, the old spring should be renewed. It is bolted to a bracket on the steering column, at the top end.

5   The parts most likely to wear are the bushes and needle roller bearings in the front suspension pivot. The bushes, liners and bearings can be pressed out of the pivot cross tube for examination; make note of the manner in which they are assembled. Each bearing has 18 needle rollers. If wear is evident, the whole assembly should be renewed, in conjunction with the pivot pin.

6   The trunnion at the base of the coil spring will wear if it is not greased regularly, as will the pivot pin which passes through it to form the pivot for the brake shoes. Both should be renewed if wear is evident; the trunnion is retained to the base of the spring by a nut, spring washer and a spring retainer cap.

7   If the stub axle bearings have any play or show signs of roughness as they are rotated, they should be drifted out of position and new replacements fitted. Refer to Chapter 5, Section 4 for the relevant procedure.

## 6   Reassembling the front suspension

1   Reassembly is accomplished by following the dismantling procedure in reverse. All moving parts and pivots should be well greased prior to reassembly; no undue force should prove necessary to replace any of the parts.

2   When relocating the front damper unit, note that the bottom mounting has two projecting lugs which must engage with recesses on the front hub casting. It may be necessary to compress the coil spring slightly before correct engagement is made.

3   Both stub axle bearings must be packed with grease; take care that none finds its way onto the brake shoes or brake drum during reassembly.

4   Care is necessary when checking the adjustment of the steering head bearings. If the bearings are too slack, the front suspension will judder, especially when the front brake is applied hard. There should be no play at the steering head bearings when the front brake is applied and the handlebars are pulled and pushed.

5   Overtight steering head bearings are equally undesirable. It is possible to unwittingly apply a loading of several tons on the head bearings by over-zealous tightening, even if the handlebars appear to turn freely. Overtight bearings will cause the machine to roll at low speeds and give generally imprecise handling. Adjustment is correct if there is no measurable play at the bearings and the handlebars swing freely from lock to lock when the machine is on the centre stand, with the front wheel off the ground. Only a very light tap on each end should cause the handlebars to swing.

## 7   Dismantling and reassembling the handlebars

1   It is rarely necessary to completely dismantle the handlebars, although occasions may arise, such as accident damage, when this is necessary.

2   Commence by following the procedure described in Section 2 of this Chapter, paragraphs 2 and 3. Then remove the headlamp unit completely by detaching the leads from the wiring harness. Release the clutch cable from the external lever under the engine and unscrew the adjuster from the cable stop.

3   Remove the right hand side cover and the lid of the air cleaner box which is retained by two screws. Detach the throttle cable from the carburettor; the cable nearest to the outer edge of the air cleaner box.

4   Remove the cover over the gear selector mechanism and

**Fig. 4.1. Front suspension and steering head**

| | | | |
|---|---|---|---|
| 1 | Damper rod | 19 | Upper rubber mounting |
| 2 | Washer | 20 | Damper trunnion |
| 3 | Spring | 21 | Conical spring |
| 4 | Damper valve pin | 22 | Washer |
| 5 | Damper piston | 23 | Discharge valve |
| 6 | Washer | 24 | Damper securing bolt - upper |
| 7 | Sleeve nut | 25 | Spring washer |
| 8 | Lower oil container | 26 | Spring washer |
| 9 | Lower rubber mounting | 27 | Damper nut - lower |
| 10 | Sleeve for rubber mounting | 28 | Plain washer |
| 11 | Valve assembly complete | 29 | Damper securing bolt - lower |
| 12 | Damper rod assembly complete | 30 | Dust cover |
| 13 | Damper main body | 31 | Lower steering head cone |
| 14 | Oil seal | 32 | Bearing cage - lower |
| 15 | Sleeve nut | 33 | Lower bearing cup |
| 16 | Packing ring | 34 | Upper bearing cup |
| 17 | Packing washer | 35 | Bearing cage - upper |
| 18 | Sleeve for upper mounting | 36 | Upper bearing sleeve nut |
| 38 | Upper bearing lock nut | 37 | Lock washer |
| 39 | Steering lock | 56 | Pin |
| 40 | Steering lock cover | 57 | Brake link |
| 41 | Pin for cover | 58 | Brake link |
| 42 | Key | 59 | Split pin |
| 43 | Nut | 60 | Nut |
| 44 | Spring washer | 61 | Brake operating lever |
| 45 | Upper spring retainer cap | 62 | Trunnion |
| 46 | Front suspension spring | 63 | Brake operating spindle |
| 47 | Nut | 64 | Brake operating lever |
| 48 | Spring washer | 65 | Split pin |
| 49 | Lower spring retainer cap | 66 | Front damper assembly complete |
| 50 | Packing | 67 | Outer sleeve |
| 51 | Plain washer | 68 | Guide bush |
| 52 | Grease nipple | 69 | Bush packing |
| 53 | Adjuster screw for brake cable | 70 | Lower bearing assembly complete |
| 54 | Nut | 71 | Upper bearing assembly complete |
| 55 | Bolt | 72 | Packing |
| | | 73 | Plain washer |

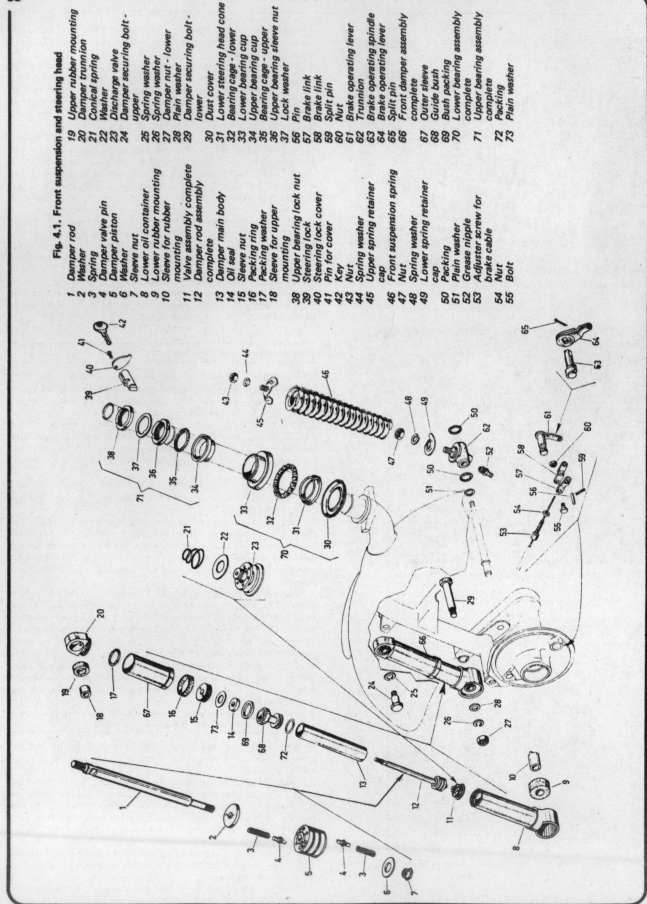

67

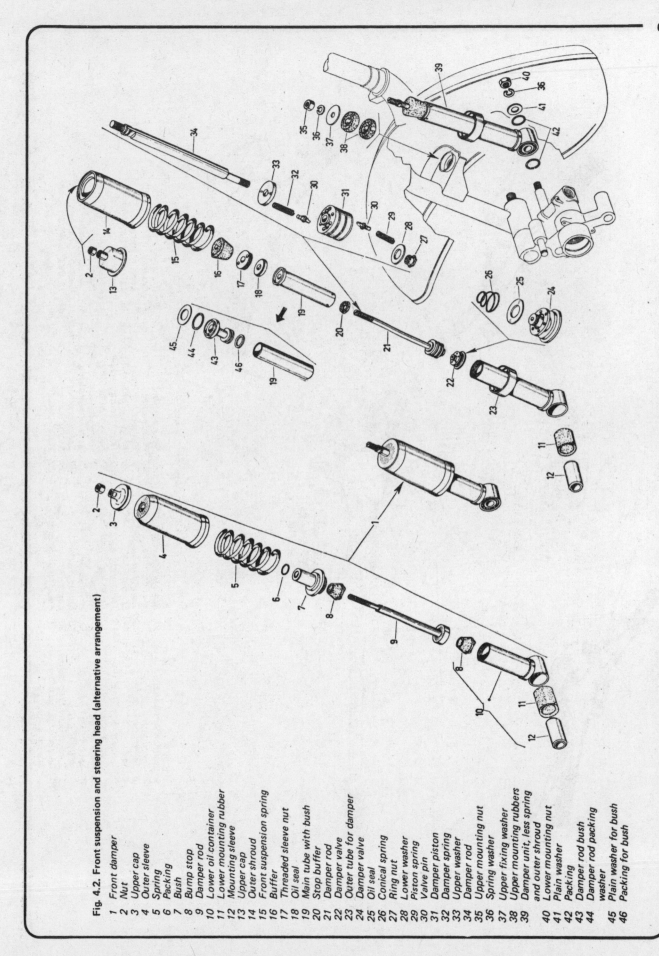

Fig. 4.2. Front suspension and steering head (alternative arrangement)

1 Front damper
2 Nut
3 Upper cap
4 Outer sleeve
5 Spring
6 Packing
7 Bush
8 Bump stop
9 Damper rod
10 Lower oil container
11 Lower mounting rubber
12 Mounting sleeve
13 Upper cap
14 Outer shroud
15 Front suspension spring
16 Buffer
17 Threaded sleeve nut
18 Oil seal
19 Main tube with bush
20 Stop buffer
21 Damper rod
22 Damper valve
23 Outer tube for damper
24 Damper valve
25 Oil seal
26 Conical spring
27 Ring nut
28 Lower washer
29 Piston spring
30 Valve pin
31 Damper piston
32 Damper spring
33 Upper washer
34 Damper rod
35 Upper mounting nut
36 Spring washer
37 Upper fixing washer
38 Upper mounting rubbers
39 Damper unit, less spring and outer shroud
40 Lower mounting nut
41 Plain washer
42 Packing
43 Damper rod bush
44 Damper rod packing washer
45 Plain washer for bush
46 Packing for bush

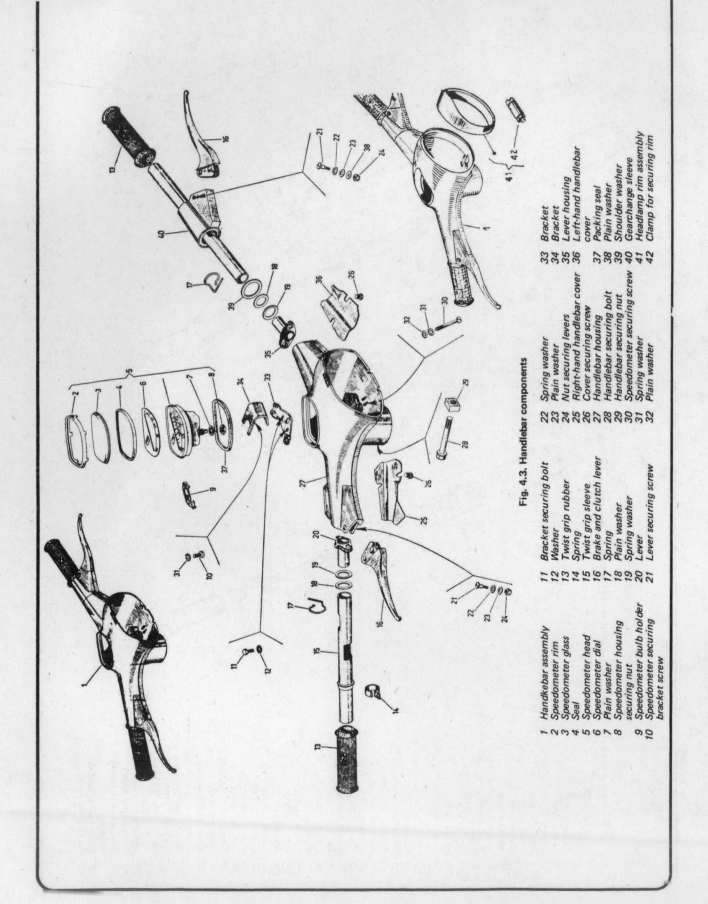

Fig. 4.3. Handlebar components

1  Handlebar assembly
2  Speedometer rim
3  Speedometer glass
4  Seal
5  Speedometer head
6  Speedometer dial
7  Plain washer
8  Speedometer housing
9  Speedometer securing nut
10 Speedometer bulb holder securing
   bracket screw
11 Bracket securing bolt
12 Washer
13 Twist grip rubber
14 Spring
15 Twist grip sleeve
16 Brake and clutch lever
17 Spring
18 Plain washer
19 Spring washer
20 Lever
21 Lever securing screw
22 Spring washer
23 Plain washer
24 Nut securing levers
25 Right-hand handlebar cover
26 Cover securing screw
27 Handlebar housing
28 Handlebar securing bolt
29 Handlebar securing nut
30 Speedometer securing screw
31 Spring washer
32 Plain washer
33 Bracket
34 Bracket
35 Lever housing
36 Left-hand handlebar cover
37 Packing seal
38 Plain washer
39 Shoulder washer
40 Gearchange sleeve
41 Headlamp rim assembly
42 Clamp for securing rim

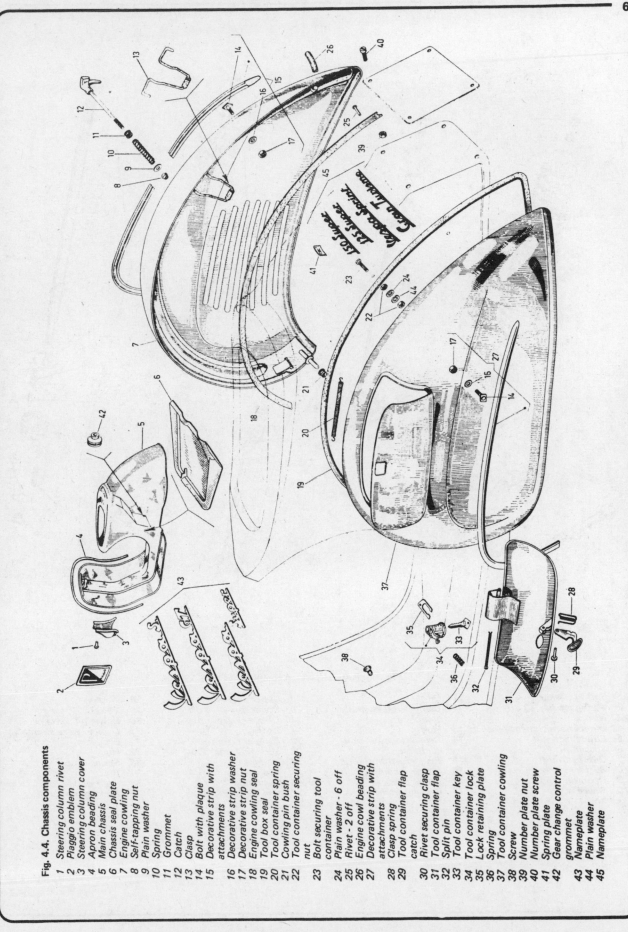

**Fig. 4.4. Chassis components**

1 Steering column rivet
2 Piaggio emblem
3 Steering column cover
4 Apron beading
5 Main chassis
6 Chassis seal plate
7 Engine cowling
8 Self-tapping nut
9 Plain washer
10 Spring
11 Grommet
12 Catch
13 Clasp
14 Bolt with plaque
15 Decorative strip with attachments
16 Decorative strip washer
17 Decorative strip nut
18 Engine cowling seal
19 Tool box seal
20 Tool container spring
21 Cowling pin bush
22 Tool container securing nut
23 Bolt securing tool container
24 Plain washer - 6 off
25 Rivet - 2 off
26 Engine cowl beading
27 Decorative strip with attachments
28 Clasp spring
29 Tool container flap catch
30 Rivet securing clasp
31 Tool container flap
32 Split pin
33 Tool container key
34 Tool container lock
35 Lock retaining plate
36 Spring
37 Tool container cowling
38 Screw
39 Number plate nut
40 Number plate screw
41 Spring plate
42 Gear change control grommet
43 Nameplate
44 Plain washer
45 Nameplate

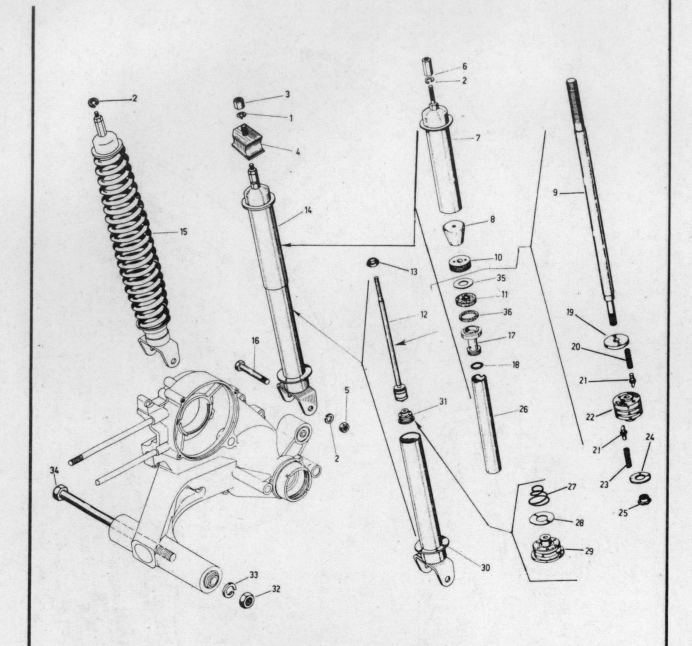

**Fig. 4.5. Rear suspension components**

| | | | | | | |
|---|---|---|---|---|---|---|
| 1 | Spring washer | 11 | Oil seal | 19 | Upper washer | 28 | Washer |
| 2 | Spring washer - 3 off | 12 | Damper rod assembly complete | 20 | Upper spring - 4 off | 29 | Valve |
| 3 | Nut | | | 21 | Damper valve spring - 2 off | 30 | Lower oil container |
| 4 | Damper attachment block | 13 | Plug | | | 31 | Damper valve assembly complete |
| 5 | Lower lock nut | 14 | Rear suspension unit complete | 22 | Damper piston | | |
| 6 | Spacer nut | | | 23 | Spring - lower - 2 off | 32 | Nut |
| 7 | Outer shroud | 15 | Rear suspension spring | 24 | Washer | 33 | Spring washer |
| 8 | Buffer | 16 | Lower mounting bolt | 25 | Cap nut | 34 | Bolt |
| 9 | Operating rod | 17 | Guide bush | 26 | Main tube | 35 | Plain washer |
| 10 | Sleeve nut | 18 | Packing washer | 27 | Conical spring | 36 | Packing washer |

detach the two control cables by unscrewing the clamping nipples. The handlebar control levers can now be removed by unscrewing the pivot bolts. Take care not to lose the anti-rattle washers. Remove the clutch and front brake cables from the levers.

5 Disconnect the barrel nuts from the handlebar ends of the cables and pull the forward portion of both the clutch and front brake cables through the handlebars into the centre so that they protrude through the headlamp aperture.

6 Raise the flap on the left hand side cover and disconnect the battery. Pull off the switch cover, take out the retaining screw and remove the switch lever block from the right hand end of the handlebars. Remove the terminal screws and the centre screw holding the switch base to the handlebars. Make a note of the arrangement of the electrical leads to aid subsequent reconnection. The cable harness can now be pulled through the handlebars and left to protrude from the headlamp aperture.

7 Remove the brackets which act as cable stops for the gear change cables and the throttle. Each is retained by two bolts; access is through the headlamp aperture. Do not lose the tensioning and Paxolin washers from behind the brackets. The throttle and gear change twist grips can now be pulled off the handlebar ends.

8 If the clamp around the base of the handlebar stem is slackened and removed, the handlebars can be lifted upward and away from the machine.

9 To reassemble the handlebars, follow the dismantling procedure in reverse. Note that when the brackets for the handlebar control cable stops are assembled, the tang of the tensioning washers must engage with the hole in the bracket. If any problems are encountered relating to the wiring of the handlebar switch, refer to the relevant wiring diagram at the end of Chapter 6 of this manual.

## 8 Chassis - examination and renovation

1 The chassis takes the form of a one-piece metal pressing and only the right and left hand side covers can be detached for renewal in the event of damage.

2 If the chassis is more badly damaged, it will be necessary to strip the machine completely, so that the bare chassis can be examined closely to ascertain the extent of the damage. Generally speaking, only very minor straightening is permissible without the application of heat. There are a few scooter repair specialists with the necessary jigs and mandrels for chassis straightening, but in most cases it is recommended that a new chassis is obtained and used, especially since there is no easy means of taking into account the effects of metal fatigue.

3 In most cases a service replacement chassis can be obtained on special order from a Vespa dealer which will provide the most economic means of handling a crash repair.

## 9 Control cables - removal and replacement

1 Although no difficulty should be experienced when removing old cables, it proves helpful to leave them in position until the new replacements have been obtained. They can then be used as a pilot to guide the replacement cables into position, by the simple expedient of attaching the new cables to their ends.

2 The most difficult cable to renew is the throttle cable which passes through the right hand side of the fuel compartment to re-appear at the horn aperture. Attach the new throttle cable outer to the handlebar end of the old cable which has first been disconnected at both ends. Keep the new cable to the right hand side of the old cable and use the latter to pull the new cable into position, when it is withdrawn from the carburettor end.

3 The clutch cable should be inserted from the rear, through the inner of the three holes at the rear of the right hand footboard until it re-appears in the horn aperture. It should protrude through the elongated slot to the left of the wiring harness.

4 The gear change cables follow a similar route to that of the clutch cable, but are inserted through the outer of the three holes in the end of the right hand footboard. They should be positioned at the extreme left of the elongated slot.

5 To replace the rear brake cable, insert the inner cable first, from immediately behind the brake pedal on the chassis, and continue feeding through until it protrudes from the underside of the machine. Slide the outer cable back over the inner cable in the reverse direction, through the centre of the three holes in the right hand footboard.

6 The choke cable operates through a guide tube in the front of the fuel compartment. The cranked end of the inner cable is attached to the choke operating knob and retained in position by a wire clip which acts also as a stop. The looped end of the cable engages with the starting device on the carburettor, inside the air cleaner box. When replacing the cable, insert the new cable complete from the front of the guide tube. Connect the cranked end to the choke knob and draw the assembly back into the guide tube. Slide the wire clip into the slot cut in the tube and pass the looped end of the cable through the grommet in the side of the chassis, adjacent to the throttle cable and wiring harness. Reconnect this end with the carburettor starting device.

## 10 Steering head lock - removal and replacement

1 A steering lock is fitted to the steering column, immediately behind the right hand side of the front apron of the chassis. The lock can be operated when the handlebars are turned to the extreme left, when the lock will engage with lugs welded on the steering column. This effectively prevents the machine from being wheeled or ridden until the lock is released.

2 If the lock malfunctions, it will have to be replaced by a new lock and key. Vespa service tool T0018111 is recommended for freeing the screwed cap which retains the lock to the chassis; the lock must then be engaged before it will pass through the housing.

3 Replace the lock by reversing the dismantling procedure.

## 11 Rear suspension - removal, examination and replacement

1 The rear suspension system is integral with the engine/gear unit and final drive; in consequence, the complete engine/gear unit must be removed as described in Chapter 1, Section 5.

2 The coil spring and the rear damper unit function in similar manner to those employed for the front suspension, although they differ in design and are not interchangeable. To check the damping action of the unit, refer to Section 5 of this Chapter, paragraphs 1 to 3.

3 Examine the coil spring as described in paragraph 4 of Section 5 of this Chapter. The spring fits over the outside of the hydraulic suspension unit and abuts against a plate at the top of the unit. The complete assembly is then held to the chassis by means of a nut and spring washer; a rubber block interposed between the chassis and the coil spring and damper unit will come away at the same time when the nut is removed. The point of attachment to the engine unit takes the form of a Silentbloc bush.

4 A Silentbloc bush system is also used for the swinging arm pivot and when wear develops in either case, renewal of the Silentbloc bushes is essential. This is not a particularly easy task because the old bushes are firmly in position. This type of repair is best entrusted to a Vespa agent who will have the necessary repair facilities. It is easy to fracture or overstress the alloy bearing housings if undue force is used to extract the old bushes.

5 Fit new pivot bolts when the Silentbloc bushes are renewed, otherwise wear of the inner bushes will be accelerated. Reassemble by reversing the dismantling procedure.

## 12 Saddle - removal and replacement

1  A saddle of the continental type, having a central suspension spring, is usually fitted as standard. In some cases, a dualseat may take its place, depending on the model and the requirements of the rider.

2  Both forms of seating arrangement bolt to a metal plate at the forward end of the petrol compartment of the chassis. The saddle pin should be greased at regular intervals to offset wear, which may otherwise give a feeling of instability whilst riding the machine.

3  When a dualseat is fitted, two mounting brackets are used, one at the front and one at the rear of the petrol compartment. The dualseat hinges on the pivot mounting of the front bracket so that it can be raised to give access to the petrol filler. The rear bracket contains a catch for locking the dualseat in the lowered position. The pivot should be greased occasionally, also the catch mechanism.

## 13 Centre stand - examination and maintenance

1  A wide legged centre stand is fitted to the Vespa scooter so that it can be parked in a stable position. The stand also facilitates the removal and replacement of either wheel in the event of a puncture.

2  It is rarely necessary to remove the centre stand unless it is damaged. It is held to the chassis by two 10 mm bolts and is retained in the retracted position by a return spring between the chassis and the stand.

3  Lubricate the stand pivots at regular intervals and make sure the return spring is in good condition. If the spring weakens or signs of wear become apparent, the spring should be renewed. If the stand drops whilst the machine is in motion, it may unseat the rider and cause an accident.

## 14 Fitting a sidecar

1  It is possible to fit a small sidecar or a delivery box to a Vespa scooter, provided the model to be fitted has been designed especially for attachment to a scooter.

2  It is strongly recommended that the fitting of a sidecar is entrusted to a scooter specialist, who has experience of sidecar alignment. The way in which the sidecar is attached is critical if the completed outfit is to handle well. An unstable outfit can prove a menace to other road users, as well as to the rider since it may react in an unexpected manner when an emergency situation occurs.

3  As a general guide, the scooter should have a lean-out of about 1½ inches, measured at handlebar level and the sidecar should have a toe-in of about the same amount. The sidecar wheel must be well forward of the rear wheel of the scooter by as much as 10 to 12 inches, and never less than 6 inches. These are only approximations; if a person experienced in sidecar fitting sets the alignment he will have his own views on the settings, based on his earlier experience.

4  Several firms have carried out work on chassis design for single seat sidecars which can be attached to a scooter, including Watsonian, Blacknell, Canterbury and Surrey. Whilst none of these sidecars is still in production, it is still possible to obtain good secondhand examples from a sidecar specialist.

## 15 Cleaning a scooter

1  After removing all surface dirt with a rag or sponge washed frequently in clean water, the application of car polish or wax will give a good finish to the machine. The plated parts should require only a wipe over with a damp rag, followed by polishing with a dry rag. If, however, corrosion has taken place, which may occur when the roads are salted during the winter, a proprietary chrome cleaner can be used.

2  The polished alloy parts will lose their sheen and oxidise slowly if they are not polished regularly. The sparing use of metal polish or a special polish such as Solvol Autosol will restore the original finish with only a few minutes labour.

3  The machine should be wiped over immediately after it has been used in the wet so that it is not garaged under damp conditions which will encourage rusting and corrosion. Remember there is little chance of water entering the control cables if they are lubricated regularly, as recommended in the Routine Maintenance Section.

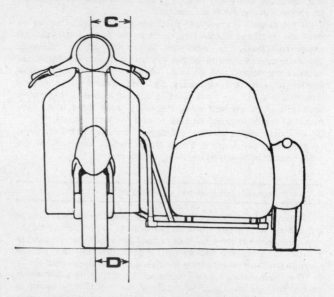

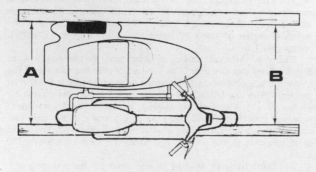

Fig. 4.6a. Aligning the sidecar wheel to the correct amount of toe-in

A – B = Toe-in

Fig. 4.6b. Setting the amount of 'lean out' by using a plumb line

C – D = Lean-out

**16 Fault diagnosis**

| Symptom | Reason/s | Remedy |
| --- | --- | --- |
| Machine veers to left or right with hands off handlebars | Incorrect wheel alignment<br>Bent forks<br>Twisted frame | Check and re-align.<br>Check and replace.<br>Check and replace. |
| Machine rolls at low speeds | Overtight steering head bearings | Slacken and re-test. |
| Machine judders when front brake is applied | Slack steering head bearings | Tighten until all play is taken up. |
| Machine pitches badly on uneven surfaces | Ineffective fork dampers<br>Ineffective rear suspension units | Check oil content.<br>Check damping action. |
| Machine wanders. Steering imprecise, rear wheel tends to hop | Worn swinging arm pivot | Dismantle and replace bushes and pivot shaft. |

# Chapter 5 Wheels, brakes and tyres

## Contents

## Specifications

### Wheels

| | | |
|---|---|---|
| Front and rear ... ... ... ... ... ... ... ... ... ... ... | Fully interchangeable | |
| Diameter | | |
|   VBC1 (Super) ... ... ... ... ... ... ... ... ... ... ... | 8 inch | |
|   All other models ... ... ... ... ... ... ... ... ... ... ... | 10 inch | |

### Tyres

| | | |
|---|---|---|
| Section | | |
|   All 90 cc and 125 cc models ... ... ... ... ... ... ... ... | 3.00 in | |
|   All other models... ... ... ... ... ... ... ... ... ... ... ... | 3.50 in | |

### Pressures, psi

| | | |
|---|---|---|
| Tyre section, in ... ... ... ... ... ... ... ... ... ... ... ... | 3.00 | 3.50 |
| Front tyre ... ... ... ... ... ... ... ... ... ... ... ... ... | 17 | 16 |
| Rear tyre, solo ... ... ... ... ... ... ... ... ... ... ... ... | 23 | 20 |
| Rear tyre, pillion passenger ... ... ... ... ... ... ... ... ... | 32 | 32 |
| Rear, with sidecar ... ... ... ... ... ... ... ... ... ... ... | — | 24 |
| Sidecar wheel ... ... ... ... ... ... ... ... ... ... ... ... | — | 16 |

## 1 General description

The Vespa scooters covered in this Manual are fitted with 8 inch diameter wheels. Each wheel carries a 3.50 section tyre which normally has a block tread pattern. This also applies to the spare wheel, mounted behind the pillion passenger.

The wheel rims are of the split type, which means that the wheel can be separated into two halves by removing the five nuts, bolts and spring washers around the periphery of the wheel. Both wheels are fully interchangeable, making it possible to carry a spare wheel as a standby in the event of a puncture. Even if a spare wheel is not available, tyre removal and replacement is greatly simplified by the split rim arrangement.

Both brakes are of the normal drum variety, with internal expanding brake shoes. The brake drum remains in position on the machine when either wheel is removed.

## 2 Front wheel - examination and renovation

1  Position the machine on the centre stand so that the front wheel is raised clear of the ground. Spin the wheel and check the rim alignment. Very small irregularities can be tolerated, but if the amount of deviation is noticeable, it is advisable to fit the spare wheel and use the existing front wheel as a spare, for emergency use only. A buckled or distorted wheel will give inferior roadholding.

2  Check whether there is any play at the wheel bearings by pulling and pushing on alternate sides of the tyre. If play is evident, the wheel bearings need renewing. Refer to Section 4 of this Chapter for the relevant information. Spin the wheel to make sure that the brakes are not binding and to ensure there is no roughness in the bearings which will also necessitate renewal.

## 3 Front brake assembly - examination, renovation and reassembly

1  Remove the front wheel by detaching the four large diameter nuts close to the wheel centre. The wheel complete with tyre and tube can then be lifted away, exposing the finned brake drum. The brake drum cover is secured by two screws; when the screws are removed, the brake drum can be lifted off along the wheel studs, exposing the brake shoes and the brake operating mechanism within the brake drum.

2  Examine the condition of the brake linings. If they are thin or worn unevenly, the brake shoes should be renewed. The linings are bonded on and cannot be supplied separately. A service exchange scheme for the renewal of brake shoes is available through a Vespa agent and will help cut costs.

3  To remove the brake shoes, detach the spring clip from the end of the common pivot shared by both shoes, then pull them apart against the return spring tension whilst easing them off the brake plate. When the brake shoes have been separated from the brake plate, the return spring can be removed.

4  Before fitting new brake shoes or replacing the originals, check that the brake operating cam operates quite smoothly. To remove the operating cam for greasing, first detach the brake cable from the end of the operating arm by slackening the cable clamp. If the split pin is withdrawn from the outer end of the operating arm, the operating arm can be pulled off the cam spindle and the cam complete with spindle withdrawn from the inside of the brake plate. Clean the spindle and its housing, then grease sparingly, before replacing and reconnecting the operating arm and cable.

5  Check the inner surface of the brake drum. The surface on which the brake shoes operate should be smooth and free from score marks or indentations, otherwise brake efficiency will be impaired. Remove all traces of brake lining dust and wipe clean with a petrol-soaked rag to remove all traces of grease and oil.

6  To reassemble the brake shoes, fit the return spring joining both shoes, and start one shoe on the pivot, which should also be very lightly greased. Push the shoe downward until there is room to engage the second shoe with the pivot and continue pressing downward with the shoes held apart fully, until they clear the operating cam and are fully home on the pivot. Replace the spring clip on the end of the pivot, refit the brake drum and the two retaining screws, then replace the front wheel, making sure the retaining bolts are tight. Check the front brake and if necessary, re-adjust by means of the cable adjuster and/or inner cable clamp on the operating arm.

7  Take great care to keep oil or grease away from the brake linings or the brake drum, during reassembly. Do not over-grease any part of the operating mechanism, or grease will eventually find its way onto the linings, giving reduced braking efficiency.

### 4  Front wheel bearings - examination and replacement

1  To gain access to the wheel bearings, it is first necessary to remove the front wheel, brake drum and brake shoes, as described in the preceding Section. When the brake shoes have been removed, the front wheel spindle can be detached by pulling off the cover from the leading link of the fork unit and then removing the hexagon-headed grease retainer and the distance piece behind it. If the front wheel stub axle is prevented from rotating, the retaining nut can be unscrewed with a ring spanner and the axle driven out of position from the outside of the bearing, using a hide-faced mallet. Access is now available to both wheel bearings within the axle housing.

2  The left hand ball journal bearing is retained by a circlip which should be removed after the oil seal which precedes it has been dislodged. There is nothing to retain the right hand ball journal bearing after the grease retainer, distance piece and axle nut have been removed. Drift both bearings outward from within the axle housing, making careful note of any spacers or shims which may be displaced so that they are replaced in the same order during reassembly.

3  Wash each bearing in a petrol/paraffin mix so that all traces of the old grease are removed. Spin each bearing in turn, checking for play or any signs of roughness as they are rotated. If there is any doubt about the condition of the bearings, they should be renewed.

4  Before fitting the new bearings, clean out the old grease from the centre of the stub axle housing and repack with new grease, leaving room for expansion. Replace any spacers or shims in their original locations, before driving the bearings back into position. Refit the retaining circlip on the left hand side, making sure it engages correctly with its groove, and a new oil seal. Tap the stub axle back into position with a hide-faced mallet, taking special care when guiding it through the centre of the new oil seal. When it is fully home, fit and tighten the retaining nut, then replace the bearing spacer and grease retaining cap.

### 5  Rear wheel - examination and renovation

1  Position the machine on the centre stand so that the rear

3.1 Brake drum is retained to the stub axle by two screws

5.3 Bearing retainer has an oil seal within the centre

6.2a Remove wire clip at pivot before shoes can be lifted off

6.2b Note how shoes overlap at pivot

6.3 Split pin must be replaced as a safeguard

7.3 Location of the stop lamp switch

8.1 Speedometer drive pinion is cut on end of stub axle

wheel is raised clear of the ground. Spin the wheel and check for run-out as described in Section 2.1 of this Chapter.

2 Check whether there is any play at the wheel bearings by pulling and pushing on alternate sides of the tyre. If play is evident, the wheel bearings need renewing. Since the rear wheel is attached direct to the end of the gearbox mainshaft, a very different procedure has to be adopted from that used in the case of the front wheel bearings. It will be necessary to remove and strip the engine/gear unit so that the bearing can be driven out of position from within the crankcase casting.

3 Note there is an oil seal located within the centre of the bearing locking ring which should be renewed each time the ring is removed and replaced. This will obviate the risk of oil leaks which would cause the already small oil content of the gearbox to fall quite dramatically.

## 6 Rear brake assembly - examination, renovation and reassembly

1 Remove the rear wheel by withdrawing the split pin through the castellated nut in the centre of the wheel. Unscrew the nut and pull the wheel off the splined driveshaft, taking care not to lose the washer on which the nut seats. The wheel will come away complete with the brake drum and driving flange.

2 Although there are certain design differences, the rear brake assembly is similar to that employed for the front wheel and has the same mode of operation. Refer to Section 3 of this Chapter, commencing at paragraph 2, for the examination, renovation and

reassembly procedure.

3 When replacing the rear wheel, make sure the castellated nut is tightened fully and that it is secured with a new split pin.

## 7 Adjusting the front and rear brakes

1 Although brake adjustment is largely a question of personal setting, there should never be sufficient slack in the control cables to cause excessive movement of the controls before the brakes are applied. The front brake lever or the rear brake pedal should never be in close proximity to the handlebars and chassis floor respectively before the brake is fully applied, otherwise there is risk it may actually touch these limits of movement during an emergency stop, and prevent full brake application.

2 As a guide, the brake should commence to apply as soon as the control is operated. Cable adjusters are provided on the front brake plate and on the underside of the engine castings, in the case of the front and rear brakes respectively. They should be turned anticlockwise to take up slack in the cable so that the brake will commence to operate earlier, or anticlockwise if the brake shows a tendency to bind when the wheel is rotated. Tighten the adjuster locknut when the adjustment is correct to ensure the adjustment cannot alter.

3 After the back brake has been adjusted, it may be necessary to re-adjust the stop lamp switch so that the stop lamp does not come on too early or too late. The switch is located close to the

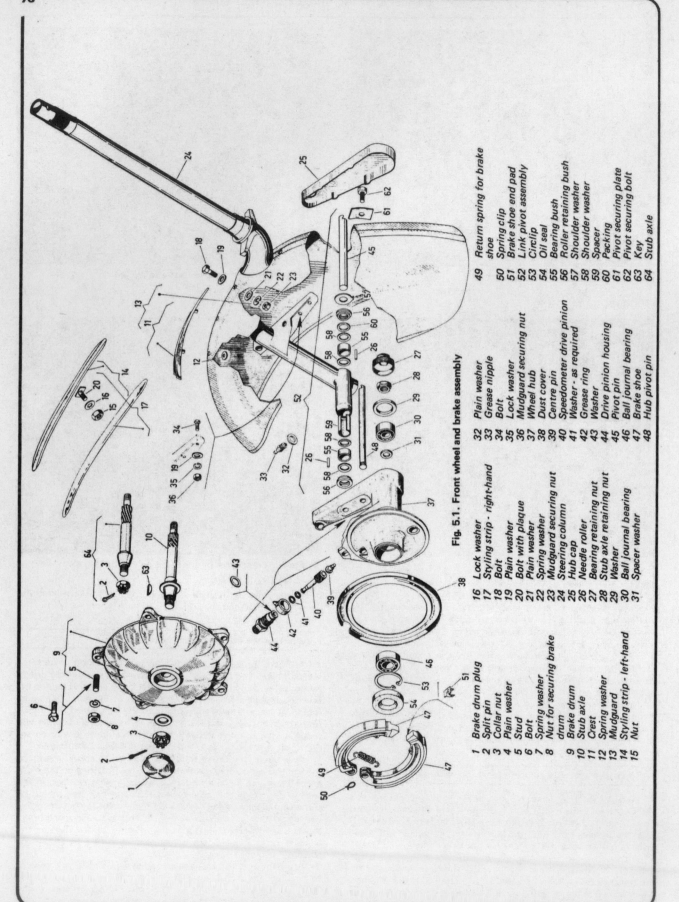

Fig. 5.1. Front wheel and brake assembly

| | | | | | | |
|---|---|---|---|---|---|---|
| 1 | Brake drum plug | 16 | Lock washer | 32 | Plain washer | 49 | Return spring for brake shoe |
| 2 | Split pin | 17 | Styling strip - right-hand | 33 | Grease nipple | 50 | Spring clip |
| 3 | Collar nut | 18 | Bolt | 34 | Bolt | 51 | Brake shoe end pad |
| 4 | Plain washer | 19 | Plain washer | 35 | Lock washer | 52 | Link pivot assembly |
| 5 | Stud | 20 | Bolt with plaque | 36 | Mudguard securing nut | 53 | Circlip |
| 6 | Bolt | 21 | Plain washer | 37 | Wheel hub | 54 | Oil seal |
| 7 | Spring washer | 22 | Spring washer | 38 | Dust cover | 55 | Bearing bush |
| 8 | Nut for securing brake drum | 23 | Mudguard securing nut | 39 | Centre pin | 56 | Roller retaining bush |
| 9 | Brake drum | 24 | Steering column | 40 | Speedometer drive pinion | 57 | Shoulder washer |
| 10 | Stub axle | 25 | Hub cap | 41 | Washer - as required | 58 | Shoulder washer |
| 11 | Crest | 26 | Needle roller | 42 | Grease ring | 59 | Spacer |
| 12 | Spring washer | 27 | Bearing retaining nut | 43 | Washer | 60 | Packing |
| 13 | Mudguard | 28 | Stub axle retaining nut | 44 | Drive pinion housing | 61 | Pivot securing plate |
| 14 | Styling strip - left-hand | 29 | Washer | 45 | Pivot pin | 62 | Pivot securing bolt |
| 15 | Nut | 30 | Ball journal bearing | 46 | Ball journal bearing | 63 | Key |
| | | 31 | Spacer washer | 47 | Brake shoe | 64 | Stub axle |
| | | | | 48 | Hub pivot pin | | |

brake pedal and can be brought nearer to or further away from the pedal by slackening the lower screw and moving the switch body in the appropriate direction. The lower mounting lug of the switch is slotted to facilitate this range of movement. Moving the switch forward makes the stop lamp operate earlier. Only a limited range of movement is available.

## 8 Speedometer drive - general

1 The speedometer drive is taken from the front wheel spindle by means of a skew-cut pinion mounted at right angles to the spindle. Skew-cut teeth formed in the end of the spindle engage with the pinion so that the drive is transmitted to the speedometer head via the customary flexible cable.

2 No attention is normally required other than the regular application of grease to the grease ring around the drive take-off during routine maintenance.

3 If the speedometer ceases to function and the drive cable is not at fault, the speedometer drive cable should be detached at its lower end by unscrewing the knurled nut at the end of the outer cable. Access to the drive pinion is obtained by unscrewing the threaded drive pinion housing which is provided with a hexagon for this purpose. When this has been removed, the drive pinion can be extracted with a pair of long nosed pliers for examination.

4 If the teeth of the drive pinion are badly worn, it will be necessary to renew this component and the front wheel spindle, with which it engages. It is bad practice to renew the drive pinion alone because the rate of wear will be greatly accelerated if the new component is permitted to engage with the worn teeth on the end of the wheel spindle.

5 It is unusual for the drive pinions to wear to the extent where they will no longer engage with each other correctly. In the comparatively few cases that do occur, failure to provide adequate lubrication is invariably the cause.

## 9 Tyres - removal and replacement

1 The usual problems which arise when it is necessary to remove and replace a tyre and tube in order to mend a puncture do not apply in the case of a scooter. Because a spare wheel is carried and both wheels are interchangeable, it is possible to change a wheel in a very short passage of time before taking to the road again.

2 If the spare wheel is not fitted, or when the wheel having the puncture has to be repaired, it is still unnecessary to use tyre levers in order to remove the tyre and tube from the wheel rim. The Vespa wheels are arranged to separate vertically so that the centre of the wheel can be pulled away from the tyre and tube, after first releasing the tyre valve from the wheel rim. When the six nuts and spring washers around the periphery of the wheel rim are removed, the wheel can be separated into two halves, making removal and replacement of the tyre and tube very easy.

3 Never run the tyres under-inflated because this will shorten their working life and may cause the valve to be ripped from the inner tube. Over-inflation is just as bad; it will affect handling and will promote uneven wear of the tyre treads. Refer to the Specifications Section of this Chapter for the recommended pressures.

## 10 Fault diagnosis - wheels, brakes and tyres

| Symptom | Reason/s | Remedy |
| --- | --- | --- |
| Handlebars oscillate at low speeds | Buckle or flat in wheel rim | Check rim alignment by spinning. Correct by fitting new wheel. |
| | Wheel nuts loose | Tighten. |
| | Tyre not straight on rim | Check tyre alignment. |
| Machine lacks power and accelerates poorly | Brakes binding | Warm brake drums provide best evidence. Re-adjust brakes. |
| Brakes grab, even when applied gently | Ends of brake shoes not chamfered | Chamfer with file. |
| | Elliptical brake drum | Lightly skim in lathe (specialist attention required). |
| Brake pull-off sluggish | Brake cam binding in housing | Free and grease. |
| | Weak brake shoe springs | Replace, if springs not displaced. |
| Middle of tyre treads wear rapidly | Tyres over-inflated | Check and re-adjust pressures. |
| Edges of tyre treads wear rapidly | Tyres under-inflated | Check and increase pressures. |

# Chapter 6 Electrical system

## Contents

## Specifications

**Battery**

| | |
|---|---|
| Type | Lead acid |
| Make | Lucas (Sportique, 232L2), Titano, SAFA, Fiamm |
| Voltage | 6 volts |
| Capacity | 7 amp/hrs to 12 amp/hrs, depending on model |
| | (90 cc and 125 cc models are not fitted with a battery) |

**Fuse**

| | |
|---|---|
| Rating | 8 amps (fitted late models only) |

**Bulbs**

| | |
|---|---|
| Main headlamp | 25/25W pre-focus |
| Parking lamp | 3W (not fitted to 90 cc or 125 cc models) |
| Tail lamp | 5W, festoon type |
| Stop lamp | 10W |
| Speedometer lamp | 0.6W, festoon type |

All bulbs rated at 6 volt

**Horn** — DC type fitted to models with batteries
AC type used on all other models, including 90 cc and 125 cc

**Earth connection** — Negative earth, models made in Italy
Positive earth, models made under licence in UK by Douglas (Sales and Service) Ltd
Check with wiring diagram in Operation and Maintenance book supplied with the machine

**A reversed battery will destroy the magnetism of the flywheel generator and render the generating system inoperative**

## 1 General description

The Vespa scooters covered by this manual are fitted with a 6 volt electrical system. The circuit comprises a crankshaft-driven AC generator, the output of which is controlled by the lighting switch. By controlling the number of generator coils in the circuit, the output from the generator can be made to match the lighting load.

A rectifier is included in the circuit so that current from the generator can be converted to DC and used for battery charging purposes.

## 2 Generator - checking the output

1 As previously mentioned in Chapter 3.2, there is no satisfactory method of checking the output from the generator without test equipment of the multi-meter type. If the performance of the generator is in any way suspect, it should be checked by either a Vespa repair specialist or an auto-electrical mechanic. Yellow lights usually indicate demagnetised magnets in the flywheel generator rotor. Refer to Chapter 3.2 for the relevant checking procedure.

## 3 Battery - examination and maintenance

1 A 6 volt battery of the lead-acid type is fitted as standard to the larger capacity models having a capacity of from 7 - 12 amp/

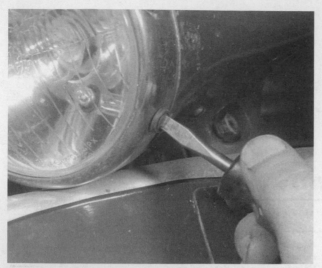

6.1 Headlamp rim and reflector unit is retained to shell by two screws

6.2a Two clips retain sealed beam bulb holder to reflector

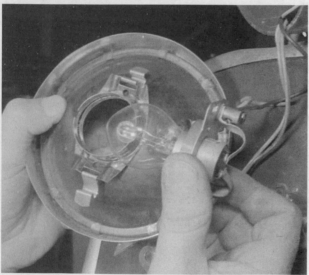

6.2b Bulb holder is push fit, has location slot to prevent mis-alignment

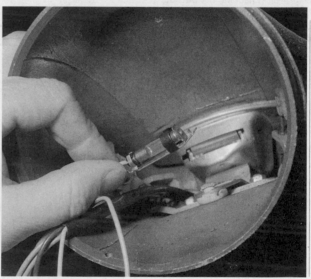

7.2 Speedometer bulb fits into side of speedometer head

9.1 Single screw retains rear lamp cover

9.2 Rear lamp has separate stop lamp and tail lamp bulbs

hours, depending on the model. A battery is not fitted to the 90 cc and 125 cc models.

2  The transparent plastic case of the battery permits the upper and lower levels of the electrolyte to be observed when the battery is lifted from its housing below the dualseat. Maintenance is normally limited to keeping the electrolyte level between the prescribed upper and lower limits and by making sure the vent pipe is not blocked. The lead plates and their separators can be seen through the transparent case, a further guide to the general condition of the battery.

3  Unless acid is spilt, as may occur if the machine falls over, the electrolyte should always be topped up with distilled water, to restore the correct level. If acid is spilt on any of the machine, it should be neutralised with an alkali such as washing soda and washed away with plenty of water, otherwise serious corrosion will occur. Top up with sulphuric acid of the correct specific gravity (1.260 - 1.280) only when spillage has occurred. Check that the vent pipe is well clear of the frame tubes or any of the other cycle parts, for obvious reasons.

## 4  Battery - charging procedure

1  The normal charging rate for a 7 amp hour battery is 0.1 amps but a more rapid charge of up to 2 amps is permissible. The higher charge rate should, if possible, be avoided since it will shorten the working life of the battery.

2  Make sure that the battery charger connections are correct, red to positive and black to negative. It is preferable to remove the battery from the machine whilst it is being charged and to remove the vent plug from each cell. When the battery is reconnected to the machine, the battery leads must be connected as shown in the wiring diagram available in the small handbook supplied with the machine. This is most important as the machine may have either a positive or a negative earth system. If the terminal connections are inadvertently reversed, the electrical system may be damaged permanently since a rectifier can be destroyed by a reverse flow of current. Furthermore, in certain circumstances, reversal polarity may demagnetise the flywheel generator rotor.

3  Periodically inspect the terminal connections to ensure corrosion is not taking place, which may eventually give rise to a high resistance connection. If corrosion has occurred, clean off the white sulphate deposit and burnish the terminals with emery cloth so that a good electrical connection can be restored. After connection, coat the terminals with a light smear of Vaseline (not grease) which will help prevent the corrosion from recurring.

## 5  Rectifier - general description

1  A rectifier of the half-wave type is fitted, which permits a one-way flow of current from the generator and therefore converts the alternating current output into direct current, which can be used for battery charging.

2  In the event of failure of the battery to maintain a fully-charged condition, it is possible that the rectifier is malfunctioning. Unfortunately there is no easy way of checking without the appropriate electronic test equipment. Provided the electrical connections have not been inadvertently transposed at the battery, a check by substitution of the correct replacement is the only practicable method of verification.

3  Later models have an 8 amp fuse incorporated in the lead from the battery to rectifier.

## 6  Headlamp - replacing bulbs

1  To remove the headlamp rim and reflector unit, detach the two small screws on the underside of the rim, located at the four o'clock and eight o'clock positions respectively. When these screws are withdrawn, the rim can be prised off the headlamp shell.

2  The main bulb is of the twin filament type to give a dipped beam facility. The bulb holder is attached to the back of the reflector unit and is retained in position by two spring clips. Remove the clips and the bulb holder will lift out of position. It is slotted to prevent misalignment on reassembly.

3  It is not necessary to refocus the headlamp when a new bulb is fitted because the bulb holder can be replaced in only one position. This obviates the risk of inadvertently reversing the position of the main and dipped beam filaments.

4  The bulb is normally rated at 25/25W, 6 volts.

5  The pilot bulb also fits within the headlamp reflector unit by means of a push fit bulb holder having bayonet fittings. The pilot bulb is rated at 3W, 6 volts. It is retained in position by the spring contact from the main bulb holder, hence it is necessary to remove the latter before the pilot bulb can be removed or replaced.

6  Variations in the form of bulb holders may occur since the type of light unit fitted has been varied on several occasions. It is always necessary, however, to remove the headlamp rim and reflector unit from the headlamp shell, before access to the bulbs is available.

## 7  Speedometer head - replacing the bulb

1  It will be necessary to remove the headlamp rim and reflector unit to gain access to the speedometer bulb, using the procedure described in the previous Section.

2  A bulb of the festoon type is used which is retained by a clip which passes into the side of the speedometer head. Only light pressure is needed to remove and replace the clip and bulb holder.

3  The speedometer bulb is rated at 6W, 6 volts.

## 8  Handlebar switch assembly

1  A switch assembly mounted on the right hand side of the handlebars controls the lighting equipment, contains the horn push and also the headlamp dip switch. The switch cover is retained by a single screw; when the cover is withdrawn, access is available to the terminal block below with the wiring connections.

2  It is rarely possible to effect a satisfactory repair if a switch malfunctions and for this reason it is advisable to renew the complete switch block as a whole, particularly if the original has seen long service.

## 9  Stop and tail lamp - replacing the bulbs

1  The tail lamp is fitted with two bulbs, a single filament bulb of the bayonet fitting type rated at 10W, 6 volts, and a festoon type bulb rated at 5W, 6 volts. The single filament bulb provides the stop lamp function, to give visual warning when the rear brake is applied. The festoon type bulb illuminates the rear number plate and also the rear of the machine when the lights are switched on.

2  To gain access to either bulb, remove the screw which retains the plastic lens cover in position. Check that the sealing gasket between the lens cover and the main body of the lamp is in good condition to prevent the ingress of water.

3  If the festoon type bulb keeps blowing, suspect either undue vibration at the rear end of the machine, or a poor or intermittent earth connection between the body of the lamp and the chassis.

## 10  Horn - location and examination

1  The horn is located behind the flared housing in the middle of the front apron of the chassis. It is of the non-adjustable type and must be renewed in the event of failure. On the 90 cc and 125 cc models having no battery, the horn is of the AC type.

2  It should be remembered that a horn in working order is a statutory requirement in the UK.

## 11 Wiring - layout and examination

1   The wiring harness is colour-coded and will correspond with the accompanying wiring diagram. Where socket connectors are used, they are designed so that reconnection can be made in the correct position only.

2   Visual inspection will show whether there are any breaks or frayed outer coverings which will give rise to short circuits. Another source of trouble may be the snap connectors and sockets, where the connector has not been pushed fully home in the outer housing.

3   Intermittent short circuits can often be traced to a chafed wire which passes through or is close to a metal component such as a frame member. Avoid tight bends in the lead or situations where a lead can become trapped between castings.

## 12 Fault diagnosis - electrical system

| Symptom | Reason/s | Remedy |
|---|---|---|
| Complete electrical failure | Short circuit | Check wiring and electrical components for short circuit. |
|  | Isolated battery | Check battery connections, also whether connections show signs of corrosion. |
| Dim lights, horn inoperative | Discharged battery | Recharge battery with battery charger and check whether alternator is giving correct output (electrical specialist). |
| Constantly 'blowing' bulbs | Vibration, poor earth connection | Check whether bulb holders are secured correctly. Check earth return or connections to frame. |

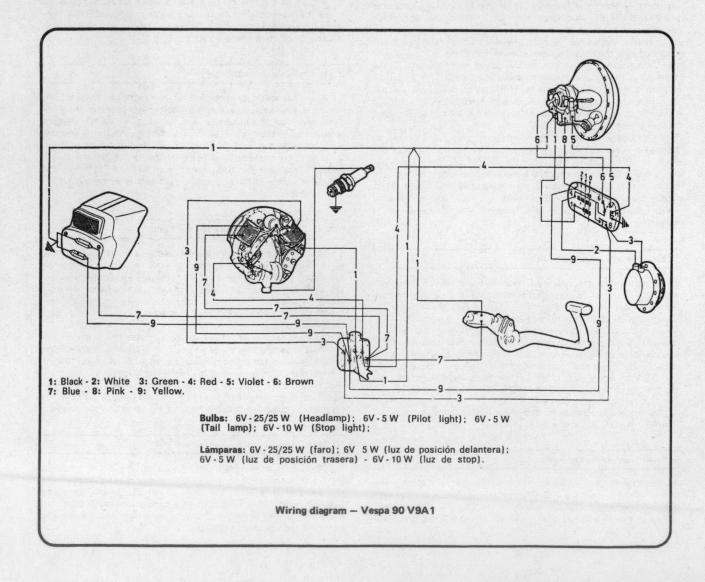

1: Black - 2: White   3: Green - 4: Red - 5: Violet - 6: Brown
7: Blue - 8: Pink - 9: Yellow.

**Bulbs:** 6V - 25/25 W  (Headlamp);  6V - 5 W  (Pilot light);  6V - 5 W (Tail lamp);  6V - 10 W  (Stop light);

**Lámparas:** 6V - 25/25 W  (faro);  6V  5 W  (luz de posición delantera); 6V - 5 W  (luz de posición trasera) - 6V - 10 W  (luz de stop).

**Wiring diagram — Vespa 90 V9A1**

1: Black - 2: Blue - 3: Yellow - 4: Red - 5: Green -
6: Brown - 7: Violet - 8: White - 9: Pink.

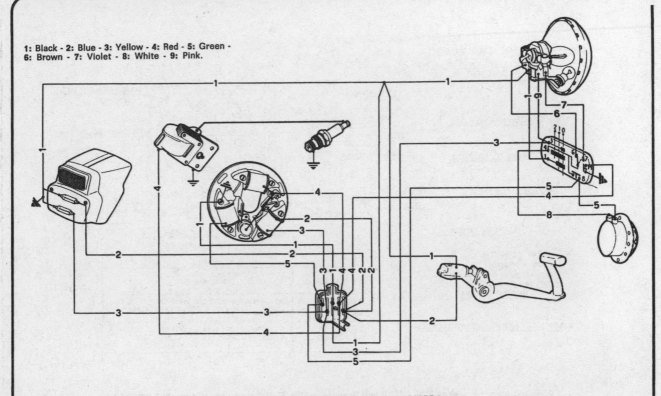

**Wiring diagram — Vespa 90 Super sport V9SS1**

1. Black - 2. Bleu - 3. Yellow - 4. Red - 5. Green - 6. Pink - 7. Violet -
8. Brown - 9. White.

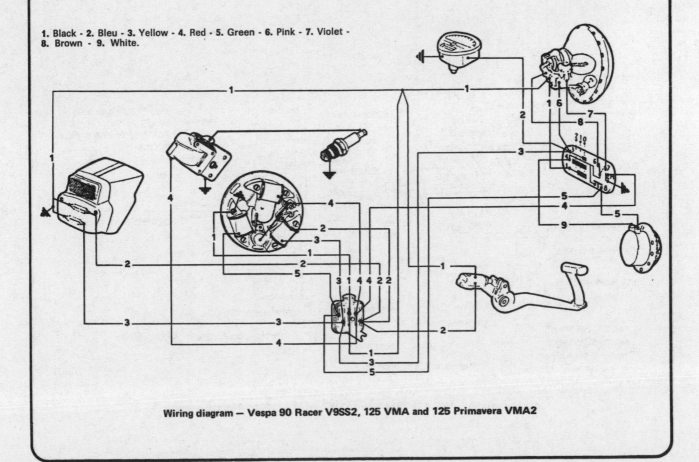

**Wiring diagram — Vespa 90 Racer V9SS2, 125 VMA and 125 Primavera VMA2**

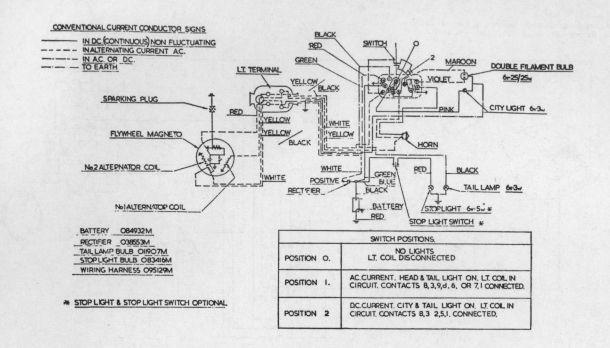

**Wiring diagram — Vespa 125 232L2 and Sportique 150 312L2 up to serial number 5AC6449**

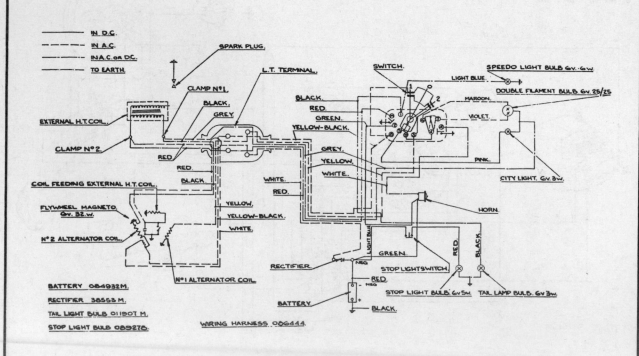

**Wiring diagram — Vespa 150 Sportique 312L2 serial number 5AC6450 onwards**

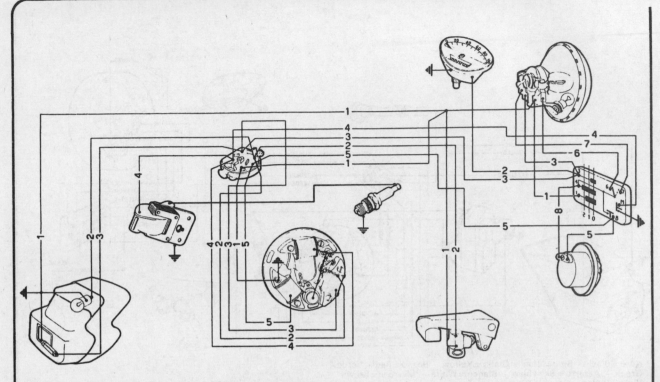

1: Black - 2: Blue - 3: Yellow - 4: Red - 5: Green - 6: Brown
7: Violet - 8: White.

**Wiring diagram — Vespa 150 Super VBC1 and 150 Sprint VLB1**

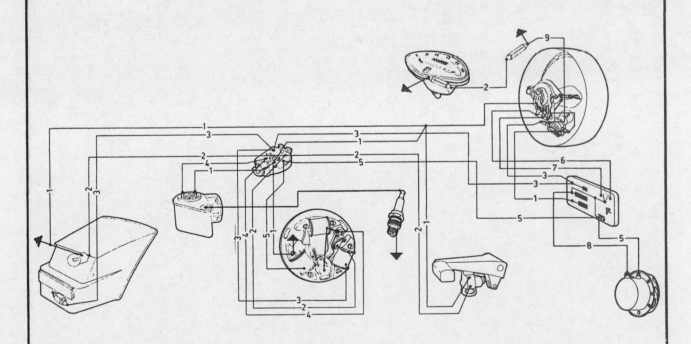

1: Black - 2: Blue - 3: Yellow   4: Red - 5: Green - 6: Violet -
7: Brown - 8: White - 9: Grey.

**Wiring diagram — Vespa 180 Rally VSD1**

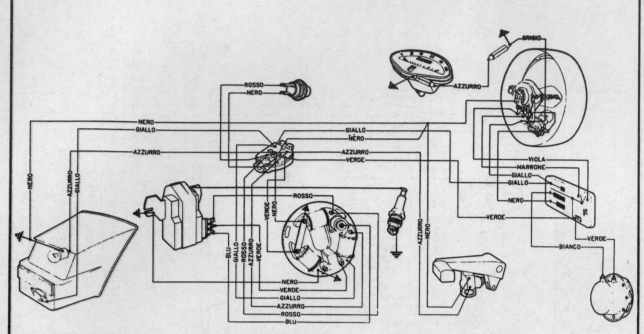

Nero=Black - Bleu=Blue - Giallo=Yellow - Rosso=Red - Verde=
Green - Azzurro=Sky-blue - Bianco=White - Marrone=Brown -
Viola=Violet - Grigio=Grey.

**Wiring diagram — Vespa 200 Rally Electronic VSE1**

# Safety first!

Professional motor mechanics are trained in safe working procedures. However enthusiastic you may be about getting on with the job in hand, do take the time to ensure that your safety is not put at risk. A moment's lack of attention can result in an accident, as can failure to observe certain elementary precautions.

There will always be new ways of having accidents, and the following points do not pretend to be a comprehensive list of all dangers; they are intended rather to make you aware of the risks and to encourage a safety-conscious approach to all work you carry out on your vehicle.

## Essential DOs and DON'Ts

**DON'T** start the engine without first ascertaining that the transmission is in neutral.

**DON'T** suddenly remove the filler cap from a hot cooling system – cover it with a cloth and release the pressure gradually first, or you may get scalded by escaping coolant.

**DON'T** attempt to drain oil until you are sure it has cooled sufficiently to avoid scalding you.

**DON'T** grasp any part of the engine, exhaust or silencer without first ascertaining that it is sufficiently cool to avoid burning you.

**DON'T** allow brake fluid or antifreeze to contact the machine's paintwork or plastic components.

**DON'T** syphon toxic liquids such as fuel, brake fluid or antifreeze by mouth, or allow them to remain on your skin.

**DON'T** inhale dust – it may be injurious to health (see *Asbestos* heading).

**DON'T** allow any spilt oil or grease to remain on the floor – wipe it up straight away, before someone slips on it.

**DON'T** use ill-fitting spanners or other tools which may slip and cause injury.

**DON'T** attempt to lift a heavy component which may be beyond your capability – get assistance.

**DON'T** rush to finish a job, or take unverified short cuts.

**DON'T** allow children or animals in or around an unattended vehicle.

**DON'T** inflate a tyre to a pressure above the recommended maximum. Apart from overstressing the carcase and wheel rim, in extreme cases the tyre may blow off forcibly.

**DO** ensure that the machine is supported securely at all times. This is especially important when the machine is blocked up to aid wheel or fork removal.

**DO** take care when attempting to slacken a stubborn nut or bolt. It is generally better to pull on a spanner, rather than push, so that if slippage occurs you fall away from the machine rather than on to it.

**DO** wear eye protection when using power tools such as drill, sander, bench grinder etc.

**DO** use a barrier cream on your hands prior to undertaking dirty jobs – it will protect your skin from infection as well as making the dirt easier to remove afterwards; but make sure your hands aren't left slippery. Note that long-term contact with used engine oil can be a health hazard.

**DO** keep loose clothing (cuffs, tie etc) and long hair well out of the way of moving mechanical parts.

**DO** remove rings, wristwatch etc, before working on the vehicle – especially the electrical system.

**DO** keep your work area tidy – it is only too easy to fall over articles left lying around.

**DO** exercise caution when compressing springs for removal or installation. Ensure that the tension is applied and released in a controlled manner, using suitable tools which preclude the possibility of the spring escaping violently.

**DO** ensure that any lifting tackle used has a safe working load rating adequate for the job.

**DO** get someone to check periodically that all is well, when working alone on the vehicle.

**DO** carry out work in a logical sequence and check that everything is correctly assembled and tightened afterwards.

**DO** remember that your vehicle's safety affects that of yourself and others. If in doubt on any point, get specialist advice.

**IF,** in spite of following these precautions, you are unfortunate enough to injure yourself, seek medical attention as soon as possible.

## Asbestos

Certain friction, insulating, sealing, and other products – such as brake linings, clutch linings, gaskets, etc – contain asbestos. *Extreme care must be taken to avoid inhalation of dust from such products since it is hazardous to health*. If in doubt, assume that they *do* contain asbestos.

## Fire

Remember at all times that petrol (gasoline) is highly flammable. Never smoke, or have any kind of naked flame around, when working on the vehicle. But the risk does not end there – a spark caused by an electrical short-circuit, by two metal surfaces contacting each other, by careless use of tools, or even by static electricity built up in your body under certain conditions, can ignite petrol vapour, which in a confined space is highly explosive.

Always disconnect the battery earth (ground) terminal before working on any part of the fuel or electrical system, and never risk spilling fuel on to a hot engine or exhaust.

It is recommended that a fire extinguisher of a type suitable for fuel and electrical fires is kept handy in the garage or workplace at all times. Never try to extinguish a fuel or electrical fire with water.

**Note:** *Any reference to a 'torch' appearing in this manual should always be taken to mean a hand-held battery-operated electric lamp or flashlight. It does **not** mean a welding/gas torch or blowlamp.*

## Fumes

Certain fumes are highly toxic and can quickly cause unconsciousness and even death if inhaled to any extent. Petrol (gasoline) vapour comes into this category, as do the vapours from certain solvents such as trichloroethylene. Any draining or pouring of such volatile fluids should be done in a well ventilated area.

When using cleaning fluids and solvents, read the instructions carefully. Never use materials from unmarked containers – they may give off poisonous vapours.

Never run the engine of a motor vehicle in an enclosed space such as a garage. Exhaust fumes contain carbon monoxide which is extremely poisonous; if you need to run the engine, always do so in the open air or at least have the rear of the vehicle outside the workplace.

## The battery

Never cause a spark, or allow a naked light, near the vehicle's battery. It will normally be giving off a certain amount of hydrogen gas, which is highly explosive.

Always disconnect the battery earth (ground) terminal before working on the fuel or electrical systems.

If possible, loosen the filler plugs or cover when charging the battery from an external source. Do not charge at an excessive rate or the battery may burst.

Take care when topping up and when carrying the battery. The acid electrolyte, even when diluted, is very corrosive and should not be allowed to contact the eyes or skin.

If you ever need to prepare electrolyte yourself, always add the acid slowly to the water, and never the other way round. Protect against splashes by wearing rubber gloves and goggles.

## Mains electricity and electrical equipment

When using an electric power tool, inspection light etc, always ensure that the appliance is correctly connected to its plug and that, where necessary, it is properly earthed (grounded). Do not use such appliances in damp conditions and, again, beware of creating a spark or applying excessive heat in the vicinity of fuel or fuel vapour. Also ensure that the appliances meet the relevant national safety standards.

## Ignition HT voltage

A severe electric shock can result from touching certain parts of the ignition system, such as the HT leads, when the engine is running or being cranked, particularly if components are damp or the insulation is defective. Where an electronic ignition system is fitted, the HT voltage is much higher and could prove fatal.

# Metric conversion tables

| Inches | Decimals | Millimetres | mm | Inches | Inches | mm |
|---|---|---|---|---|---|---|
| | | | **Millimetres to Inches** | | **Inches to Millimetres** | |
| 1/64 | 0.015625 | 0.3969 | 0.01 | 0.00039 | 0.001 | 0.0254 |
| 1/32 | 0.03125 | 0.7937 | 0.02 | 0.00079 | 0.002 | 0.0508 |
| 3/64 | 0.046875 | 1.1906 | 0.03 | 0.00118 | 0.003 | 0.0762 |
| 1/16 | 0.0625 | 1.5875 | 0.04 | 0.00157 | 0.004 | 0.1016 |
| 5/64 | 0.078125 | 1.9844 | 0.05 | 0.00197 | 0.005 | 0.1270 |
| 3/32 | 0.09375 | 2.3812 | 0.06 | 0.00236 | 0.006 | 0.1524 |
| 7/64 | 0.109375 | 2.7781 | 0.07 | 0.00276 | 0.007 | 0.1778 |
| 1/8 | 0.125 | 3.1750 | 0.08 | 0.00315 | 0.008 | 0.2032 |
| 9/64 | 0.140625 | 3.5719 | 0.09 | 0.00354 | 0.009 | 0.2286 |
| 5/32 | 0.15625 | 3.9687 | 0.1 | 0.00394 | 0.01 | 0.254 |
| 11/64 | 0.171875 | 4.3656 | 0.2 | 0.00787 | 0.02 | 0.508 |
| 3/16 | 0.1875 | 4.7625 | 0.3 | 0.01181 | 0.03 | 0.762 |
| 13/64 | 0.203125 | 5.1594 | 0.4 | 0.01575 | 0.04 | 1.016 |
| 7/32 | 0.21875 | 5.5562 | 0.5 | 0.01969 | 0.05 | 1.270 |
| 15/64 | 0.234375 | 5.9531 | 0.6 | 0.02362 | 0.06 | 1.524 |
| 1/4 | 0.25 | 6.3500 | 0.7 | 0.02756 | 0.07 | 1.778 |
| 17/64 | 0.265625 | 6.7469 | 0.8 | 0.03150 | 0.08 | 2.032 |
| 9/32 | 0.28125 | 7.1437 | 0.9 | 0.03543 | 0.09 | 2.286 |
| 19/64 | 0.296875 | 7.5406 | 1 | 0.03937 | 0.1 | 2.54 |
| 5, 16 | 0.3125 | 7.9375 | 2 | 0.07874 | 0.2 | 5.08 |
| 21/64 | 0.328125 | 8.3344 | 3 | 0.11811 | 0.3 | 7.62 |
| 11/32 | 0.34375 | 8.7312 | 4 | 0.15748 | 0.4 | 10.16 |
| 23/64 | 0.359375 | 9.1281 | 5 | 0.19685 | 0.5 | 12.70 |
| 3/8 | 0.375 | 9.5250 | 6 | 0.23622 | 0.6 | 15.24 |
| 25/64 | 0.390625 | 9.9219 | 7 | 0.27559 | 0.7 | 17.78 |
| 13/32 | 0.40625 | 10.3187 | 8 | 0.31496 | 0.8 | 20.32 |
| 27/64 | 0.421875 | 10.7156 | 9 | 0.35433 | 0.9 | 22.86 |
| 7/16 | 0.4375 | 11.1125 | 10 | 0.39370 | 1 | 25.4 |
| 29/64 | 0.453125 | 11.5094 | 11 | 0.43307 | 2 | 50.8 |
| 15/32 | 0.46875 | 11.9062 | 12 | 0.47244 | 3 | 76.2 |
| 31/64 | 0.484375 | 12.3031 | 13 | 0.51181 | 4 | 101.6 |
| 1/2 | 0.5 | 12.7000 | 14 | 0.55118 | 5 | 127.0 |
| 33/64 | 0.515625 | 13.0969 | 15 | 0.59055 | 6 | 152.4 |
| 17/32 | 0.53125 | 13.4937 | 16 | 0.62992 | 7 | 177.8 |
| 35/64 | 0.546875 | 13.8906 | 17 | 0.66929 | 8 | 203.2 |
| 9/16 | 0.5625 | 14.2875 | 18 | 0.70866 | 9 | 228.6 |
| 37/64 | 0.578125 | 14.6844 | 19 | 0.74803 | 10 | 254.0 |
| 19/32 | 0.59375 | 15.0812 | 20 | 0.78740 | 11 | 279.4 |
| 39/64 | 0.609375 | 15.4781 | 21 | 0.82677 | 12 | 304.8 |
| 5/8 | 0.625 | 15.8750 | 22 | 0.86614 | 13 | 330.2 |
| 41/64 | 0.640625 | 16.2719 | 23 | 0.90551 | 14 | 355.6 |
| 21/32 | 0.65625 | 16.6687 | 24 | 0.94488 | 15 | 381.0 |
| 43/64 | 0.671875 | 17.0656 | 25 | 0.98425 | 16 | 406.4 |
| 11/16 | 0.6875 | 17.4625 | 26 | 1.02362 | 17 | 431.8 |
| 45/64 | 0.703125 | 17.8594 | 27 | 1.06299 | 18 | 457.2 |
| 23/32 | 0.71875 | 18.2562 | 28 | 1.10236 | 19 | 482.6 |
| 47/64 | 0.734375 | 18.6531 | 29 | 1.14173 | 20 | 508.0 |
| 3/4 | 0.75 | 19.0500 | 30 | 1.18110 | 21 | 533.4 |
| 49/64 | 0.765625 | 19.4469 | 31 | 1.22047 | 22 | 558.8 |
| 25/32 | 0.78125 | 19.8437 | 32 | 1.25984 | 23 | 584.2 |
| 51/64 | 0.796875 | 20.2406 | 33 | 1.29921 | 24 | 609.6 |
| 13/16 | 0.8125 | 20.6375 | 34 | 1.33858 | 25 | 635.0 |
| 53/64 | 0.828125 | 21.0344 | 35 | 1.37795 | 26 | 660.4 |
| 27/32 | 0.84375 | 21.4312 | 36 | 1.41732 | 27 | 685.8 |
| 55/64 | 0.859375 | 21.8281 | 37 | 1.4567 | 28 | 711.2 |
| 7/8 | 0.875 | 22.2250 | 38 | 1.4961 | 29 | 736.6 |
| 57/64 | 0.890625 | 22.6219 | 39 | 1.5354 | 30 | 762.0 |
| 29/32 | 0.90625 | 23.0187 | 40 | 1.5748 | 31 | 787.4 |
| 59/64 | 0.921875 | 23.4156 | 41 | 1.6142 | 32 | 812.8 |
| 15/16 | 0.9375 | 23.8125 | 42 | 1.6535 | 33 | 838.2 |
| 61/64 | 0.953125 | 24.2094 | 43 | 1.6929 | 34 | 863.6 |
| 31/32 | 0.96875 | 24.6062 | 44 | 1.7323 | 35 | 889.0 |
| 63/64 | 0.984375 | 25.0031 | 45 | 1.7717 | 36 | 914.4 |

# Conversion factors

## Length (distance)

| | | | | | |
|---|---|---|---|---|---|
| Inches (in) | X | 25.4 | = Millimetres (mm) | X 0.0394 | = Inches (in) |
| Feet (ft) | X | 0.305 | = Metres (m) | X 3.281 | = Feet (ft) |
| Miles | X | 1.609 | = Kilometres (km) | X 0.621 | = Miles |

## Volume (capacity)

| | | | | | |
|---|---|---|---|---|---|
| Cubic inches (cu in; in³) | X | 16.387 | = Cubic centimetres (cc; cm³) | X 0.061 | = Cubic inches (cu in; in³) |
| Imperial pints (Imp pt) | X | 0.568 | = Litres (l) | X 1.76 | = Imperial pints (Imp pt) |
| Imperial quarts (Imp qt) | X | 1.137 | = Litres (l) | X 0.88 | = Imperial quarts (Imp qt) |
| Imperial quarts (Imp qt) | X | 1.201 | = US quarts (US qt) | X 0.833 | = Imperial quarts (Imp qt) |
| US quarts (US qt) | X | 0.946 | = Litres (l) | X 1.057 | = US quarts (US qt) |
| Imperial gallons (Imp gal) | X | 4.546 | = Litres (l) | X 0.22 | = Imperial gallons (Imp gal) |
| Imperial gallons (Imp gal) | X | 1.201 | = US gallons (US gal) | X 0.833 | = Imperial gallons (Imp gal) |
| US gallons (US gal) | X | 3.785 | = Litres (l) | X 0.264 | = US gallons (US gal) |

## Mass (weight)

| | | | | | |
|---|---|---|---|---|---|
| Ounces (oz) | X | 28.35 | = Grams (g) | X 0.035 | = Ounces (oz) |
| Pounds (lb) | X | 0.454 | = Kilograms (kg) | X 2.205 | = Pounds (lb) |

## Force

| | | | | | |
|---|---|---|---|---|---|
| Ounces-force (ozf; oz) | X | 0.278 | = Newtons (N) | X 3.6 | = Ounces-force (ozf; oz) |
| Pounds-force (lbf; lb) | X | 4.448 | = Newtons (N) | X 0.225 | = Pounds-force (lbf; lb) |
| Newtons (N) | X | 0.1 | = Kilograms-force (kgf; kg) | X 9.81 | = Newtons (N) |

## Pressure

| | | | | | |
|---|---|---|---|---|---|
| Pounds-force per square inch (psi; lbf/in²; lb/in²) | X | 0.070 | = Kilograms-force per square centimetre (kgf/cm²; kg/cm²) | X 14.223 | = Pounds-force per square inch (psi; lbf/in²; lb/in²) |
| Pounds-force per square inch (psi; lbf/in²; lb/in²) | X | 0.068 | = Atmospheres (atm) | X 14.696 | = Pounds-force per square inch (psi; lbf/in²; lb/in²) |
| Pounds-force per square inch (psi; lbf/in²; lb/in²) | X | 0.069 | = Bars | X 14.5 | = Pounds-force per square inch (psi; lbf/in²; lb/in²) |
| Pounds-force per square inch (psi; lbf/in²; lb/in²) | X | 6.895 | = Kilopascals (kPa) | X 0.145 | = Pounds-force per square inch (psi; lbf/in²; lb/in²) |
| Kilopascals (kPa) | X | 0.01 | = Kilograms-force per square centimetre (kgf/cm²; kg/cm²) | X 98.1 | = Kilopascals (kPa) |
| Millibar (mbar) | X | 100 | = Pascals (Pa) | X 0.01 | = Millibar (mbar) |
| Millibar (mbar) | X | 0.0145 | = Pounds-force per square inch (psi; lbf/in²; lb/in²) | X 68.947 | = Millibar (mbar) |
| Millibar (mbar) | X | 0.75 | = Millimetres of mercury (mmHg) | X 1.333 | = Millibar (mbar) |
| Millibar (mbar) | X | 0.401 | = Inches of water (inH₂O) | X 2.491 | = Millibar (mbar) |
| Millimetres of mercury (mmHg) | X | 0.535 | = Inches of water (inH₂O) | X 1.868 | = Millimetres of mercury (mmHg) |
| Inches of water (inH₂O) | X | 0.036 | = Pounds-force per square inch (psi; lbf/in²; lb/in²) | X 27.68 | = Inches of water (inH₂O) |

## Torque (moment of force)

| | | | | | |
|---|---|---|---|---|---|
| Pounds-force inches (lbf in; lb in) | X | 1.152 | = Kilograms-force centimetre (kgf cm; kg cm) | X 0.868 | = Pounds-force inches (lbf in; lb in) |
| Pounds-force inches (lbf in; lb in) | X | 0.113 | = Newton metres (Nm) | X 8.85 | = Pounds-force inches (lbf in; lb in) |
| Pounds-force inches (lbf in; lb in) | X | 0.083 | = Pounds-force feet (lbf ft; lb ft) | X 12 | = Pounds-force inches (lbf in; lb in) |
| Pounds-force feet (lbf ft; lb ft) | X | 0.138 | = Kilograms-force metres (kgf m; kg m) | X 7.233 | = Pounds-force feet (lbf ft; lb ft) |
| Pounds-force feet (lbf ft; lb ft) | X | 1.356 | = Newton metres (Nm) | X 0.738 | = Pounds-force feet (lbf ft; lb ft) |
| Newton metres (Nm) | X | 0.102 | = Kilograms-force metres (kgf m; kg m) | X 9.804 | = Newton metres (Nm) |

## Power

| | | | | | |
|---|---|---|---|---|---|
| Horsepower (hp) | X | 745.7 | = Watts (W) | X 0.0013 | = Horsepower (hp) |

## Velocity (speed)

| | | | | | |
|---|---|---|---|---|---|
| Miles per hour (miles/hr; mph) | X | 1.609 | = Kilometres per hour (km/hr; kph) | X 0.621 | = Miles per hour (miles/hr; mph) |

## Fuel consumption*

| | | | | | |
|---|---|---|---|---|---|
| Miles per gallon, Imperial (mpg) | X | 0.354 | = Kilometres per litre (km/l) | X 2.825 | = Miles per gallon, Imperial (mpg) |
| Miles per gallon, US (mpg) | X | 0.425 | = Kilometres per litre (km/l) | X 2.352 | = Miles per gallon, US (mpg) |

## Temperature

Degrees Fahrenheit = (°C x 1.8) + 32          Degrees Celsius (Degrees Centigrade; °C) = (°F - 32) x 0.56

*It is common practice to convert from miles per gallon (mpg) to litres/100 kilometres (l/100km), where mpg (Imperial) x l/100 km = 282 and mpg (US) x l/100 km = 235

# English/American terminology

Because this book has been written in England, British English component names, phrases and spellings have been used throughout. American English usage is quite often different and whereas normally no confusion should occur, a list of equivalent terminology is given below.

| English | American | English | American |
|---|---|---|---|
| Air filter | Air cleaner | Number plate | License plate |
| Alignment (headlamp) | Aim | Output or layshaft | Countershaft |
| Allen screw/key | Socket screw/wrench | Panniers | Side cases |
| Anticlockwise | Counterclockwise | Paraffin | Kerosene |
| Bottom/top gear | Low/high gear | Petrol | Gasoline |
| Bottom/top yoke | Bottom/top triple clamp | Petrol/fuel tank | Gas tank |
| Bush | Bushing | Pinking | Pinging |
| Carburettor | Carburetor | Rear suspension unit | Rear shock absorber |
| Catch | Latch | Rocker cover | Valve cover |
| Circlip | Snap ring | Selector | Shifter |
| Clutch drum | Clutch housing | Self-locking pliers | Vise-grips |
| Dip switch | Dimmer switch | Side or parking lamp | Parking or auxiliary light |
| Disulphide | Disulfide | Side or prop stand | Kick stand |
| Dynamo | DC generator | Silencer | Muffler |
| Earth | Ground | Spanner | Wrench |
| End float | End play | Split pin | Cotter pin |
| Engineer's blue | Machinist's dye | Stanchion | Tube |
| Exhaust pipe | Header | Sulphuric | Sulfuric |
| Fault diagnosis | Trouble shooting | Sump | Oil pan |
| Float chamber | Float bowl | Swinging arm | Swingarm |
| Footrest | Footpeg | Tab washer | Lock washer |
| Fuel/petrol tap | Petcock | Top box | Trunk |
| Gaiter | Boot | Torch | Flashlight |
| Gearbox | Transmission | Two/four stroke | Two/four cycle |
| Gearchange | Shift | Tyre | Tire |
| Gudgeon pin | Wrist/piston pin | Valve collar | Valve retainer |
| Indicator | Turn signal | Valve collets | Valve cotters |
| Inlet | Intake | Vice | Vise |
| Input shaft or mainshaft | Mainshaft | Wheel spindle | Axle |
| Kickstart | Kickstarter | White spirit | Stoddard solvent |
| Lower leg | Slider | Windscreen | Windshield |
| Mudguard | Fender | | |

# Index